Urban Biodiversity and Built Environment

Jing Gan

Urban Biodiversity and Built Environment

Case Study of Shanghai

同濟大學出版社
TONGJI UNIVERSITY PRESS

Springer

Jing Gan
Department of Urban Planning
Tongji University
Shanghai, China

ISBN 978-981-16-0948-0 ISBN 978-981-16-0949-7 (eBook)
https://doi.org/10.1007/978-981-16-0949-7

Jointly published with Tongji University Press
The print edition is not for sale in China (Mainland). Customers from China (Mainland) please order the print book from: Tongji University Press.

This Springer imprint is published by the registered company Springer Nature Singapore Pte Ltd.
The registered company address is: 152 Beach Road, #21-01/04 Gateway East, Singapore 189721, Singapore

Foreword

At the end of 1997, when I drafted the *Declaration of Urban Planners for the 21st century*, it was just when cities across the world were at the turning point to enter a new urban century. When I proposed the "three major Harmonies" as the goal for urban planners, I seriously listed "harmony between people and nature" at the top. The other two goals are "harmony between people in the cities" and "harmony between future and past".

Faced with the increasingly serious problems and challenges of urban ecological environment, in the new century, I gradually integrate my idea of city-being (city as living being) into all aspects of urban ecological planning and design. I had supervised a couple of master students who studied the ecological dimensions of "energy", "water", "material", "air flow" and "land" in their thesis research, but had left the last dimension "biodiversity" at that time, as I never found a suitable student to complete it.

I had also seen many attempts in pilot eco-cities such as Harmarby, Sweden. These world-well known "eco cities" have actually made achievements only in the three aspects of "energy", "water" and "material", but there is still a critical distance from the "bio" aspect of "Eco".

In 2010, when GAN Jing started her doctoral program many years after her master's degree, I suggested her to choose the relationship between urban biodiversity and built environment as the research topic, examine the interactive relationship between city and natural environment, and give full play to the positive role of urban planning and design in conserving and enhancing biodiversity in cities, to supplement the lack of eco-city research dimension. In the research process, she overcame the interdisciplinary barriers and the lack of basic data, studied the relevant literature of urban ecology and urban biology, obtained first-hand data with long-term field study, and completed satisfactory academic results.

This book is based on Dr. GAN's doctoral dissertation. Aiming at the frontier problem of the lack of biological habitat system in China's urban planning and construction, it studies the relationship between urban biodiversity and built environment from the perspective of urban planning and design. Through empirical study, it discusses the impact of urban built environment of different scales on urban biodiversity with avian diversity as a focus. It can provide a theoretical basis for the preparation, implementation, and management of urban ecological planning and design, to protect and enhance urban biodiversity in practice, and to promote the development of high-density living environment and natural habitat in the limited land resources in China. The book can also provide basis for the optimization of the urban built environment.

In the critical stage of China's urbanization spanning 60%, urban development in the new era takes the construction of ecological civilization as the goal orientation and the human-nature community of shared life as the guiding ideology. As an important natural resource of a city, the value of urban biodiversity should be recognized and respected, and should not be sacrificed in the process of comprehensive benefits evaluation of urban development. We hope that through more interdisciplinary research and practice, we can create a biodiversity-friendly natural environment in high-density cities, and provide citizens with visible, audible, and sensible natural ecological well-being.

Shanghai, China

Prof. Dr. Zhiqiang Wu
Member of Chinese Academy of Engineering
Member of German Academy of Science and Engineering (acatech)
Member of Royal Swedish Academy of Engineering and Science (IVA)
Honorary Fellow of American Institute of Architects (Hon. FAIA)

Acknowledgements

This book is based on my doctoral dissertation "The relationship between urban biodiversity and urban built environment". There is an instinctive bond between human beings and other living systems. China's national strategy of ecological civilization and the concept of "Building a Shared Future for All Life on Earth" provide a new basis and direction for how to protect natural resources in the process of rapid urbanization and urban development. The conservation and enhancement of biodiversity in the built-up environment of high-density cities is one of the key issues facing urban sustainable development. The integration of biodiversity perspective into urban planning and design is also an area that needs to be further explored, especially in the study of eco-city development, which can provide more reference for the future development of urban ecological transformation.

The research of urban biodiversity has become one of the hot topics in the international urbanization and eco-city research agenda since the 1990s. However, the research and practice of urban planning in China is still in its infancy, with a lack of basic data and also little professional knowledge transformation and connection between urban ecology, biological theory and urban planning practice. At the end of 2011, with the encouragement of my supervisor, Prof. Dr. Zhiqiang Wu, I chose this interdisciplinary field with enthusiasm and satisfaction to complete my doctoral dissertation. During the research process, every small step has been helped by many people. When this book is about to be published, there are too many thanks to express.

First of all, I would like to thank my supervisor tutor Prof. Dr. Zhiqiang Wu. Since I met Prof. Wu in the autumn of 1999, his profound knowledge, forward-looking thinking, keen perspective and curiosity to explore the nature of things have had a profound impact on me. During the whole process of research, I once felt that I couldn't stick to it. It was Prof. Wu's constant encouragement that made me regain my confidence and discover my potential again and again. When my research entered the bottleneck period, it was also Prof. Wu's timely guidance that pulled me back to the core issue of planning research. Also thanks to Prof. WU for the preface he wrote for this book, which makes the value and significance of the book better presented.

Thanks to Prof. Qingji Shen from the Department of Urban Planning, Tongji University, for providing a lot of opinions and suggestions for this study, which helped me to promote and improve the research results, and encouraged me to express my academic views when I was not confident in the research process at the beginning.

Thanks to Assoc. Prof. Guangpu Guo from the School of Life Science and Technology, Tongji University, with the help of Mr. Guo and his team, I was able to collect the basic data of this research. Mr. Guo's heartfelt love for nature and biology and his efforts to promote nature education also inspired me to continue my research in this field.

Thanks to Prof. Deshun Zhang from the Department of Landscape Architecture, Tongji University, for his guidance to the botany part of this study, especially in the aspects of vegetation pattern index selection, research method design, variable calculation, etc.

Thanks to Profs. Xiangming Chen and Joan Morrison of Trinity College in the US. Professor Chen invited me to Trinity as Henry Luce visiting scholar in 2014 and provided me with good funding conditions, which made it easier for me to read a large number of English literature and accumulate knowledge reserves in this field. Professor Morrison took me to conduct my first bird survey in Hartford in early spring 2014, which laid the foundation for empirical research after I returned to China.

Thanks to Prof. Jürgen Breuste-the President of the Society for Urban Ecology (SURE). The SURE summer school 2016 organized by Profs. Jürgen Breuste and Junxiang Li in Shanghai deepened my understanding of urban biodiversity research. The following exchange visits and academic discussions and the co-authored article with Prof. Breuste in the following years have deepened my original research results and inspired my further research.

Thanks to the National Natural Science Foundation of China. This research is supported by two NSFC projects (Grant No. 51978477, 51408426). The comments and suggestions from the anonymous reviewers not only strengthened my research determination, but also helped me constantly revise and clarify the research content. It is also under the continuous support by NSFC and the seed projects of the Key Laboratory of high density human settlements ecology and energy conservation of Tongji University, that the research of this study can be carried out smoothly.

Thanks to Prof. Ruolin Zhu, the President of Shanghai Pudong New Area Planning Association, Prof. Xiangrong Wang and Zhijun Ma from Fudan University, Profs. Shangwu Zhang and Nankai Xia from Tongji University for their comments on my doctoral dissertation. Thanks to Profs. Songling Tao, Zhenwei Peng, Guiqing Yang, Xiaodong Song, Jieming Zhu, Jian Zhuo, Huizhi Geng, Song Liu, Lan Wang, Xinyi Niu, Asso.Profs. Wang Yajuan, Xueyan Yao, Qing Li, Dr. Yuelai Liu, Dr. Lei Pang and Ms. Jing Cao for their concern and encouragement.

Due to the lack of basic data, many of my friends were "disturbed" by me and offered selfless help. I would like to thank Mr. Weijie Chen and Mr. Yao Huang from Shanghai Pudong New Area Planning and Design Institute, Ms. Qiuxia Gong, Chief Planner of Qiantan International Business District, Ms. Liping Chen from Pudong New Area Environmental Protection Bureau, Mr. Zhitong Zhang and Ms. Yuyi Liu from Shanghai Wildlife Conservation Management Station, Mr. YeYe Zhang from Dongchang middle school affiliated to East China Normal University, and Mr. Yongxi Mao from Pudong Jiaozhong junior middle school for helping with the basic data. Thanks to Di Zheng, Yan Wu, Ying Zhao, Qing Wang and Lei Yang for their help. Thanks to Yiran Song, Yuchen Dai, Zikun Chen, Pengcheng Jiang, Xiling Li, Wenxuan Guo, Ling Wang, Ling Leng, Xiaokai Wu, Jiankang Wu and other students, who participated in the bird survey in the blocks along Century Avenue, and Jing Yang, Wenwan Xie, Jinjin Guan, Dawei Fu and Xiaojie Wang who participated in the vegetation survey. Thanks to Lingyu Kong, Heng Zhang, Changfeng Sun, Hongbing Zhou and Lei Liu for their help in the use of various analysis software.

Thanks to my family for their support throughout the research. Thanks to my husband Yuxin Ding, who has always encouraged me with a good temper when I faced difficulties in this research. He always believes that I can achieve excellent results. Thank you for your understanding and support. Thanks to my dearest mother and parents in law. It's a pity that my father passed away in the autumn of 2015. Hope my dear daddy in the heaven will be proud of me to see my book publish.

Thanks to Dai Jiang, Deputy Director in Chief of Tongji University Press, Xiaoli Zhu and Jialin (Crisie) Yuan for their hard work in publishing the book.

Several of the chapters draw from earlier published article in *Urban Planning International*, *Planners*, *Urban Development Studies*, *Shanghai Urban Planning*, *Landscape Architecture*, *Journal of Chinese Urban Forestry*, and Springer's book *Making green cities: Concept, Challenges and Practices*, but have been modified in writing this book. There are still many contents in the field of "Urban biodiversity and built environment" that need to be further explored. At the time of the completion of the book, new research has started, and practical projects based on research results are also being carried out. I hope that more achievements can be presented to readers in the future.

The book is supported by the National Natural Science Foundation of China (Grant No. 51978477, 51408426)

Shanghai, China Jing Gan
Spring 2021

Contents

List of Figures

List of Tables

Chapter 1
Introduction

1.1 Significance to Integrate Biodiversity Protection and Enhancement into Urban Planning and Design

1.1.1 Human Activities Are Causing Increasing Damage to Biodiversity

Amid surging population growth and ever-accelerating progress of science and technology, humans have created splendid civilizations, but also have taken control of the global ecosystem. Humans' over-utilization of biological resources and land over the years has triggered a series of problems such as the loss of wildlife habitat, environmental pollution, etc., and a species extinction crisis is underway across the planet, presenting a great threat to its ecological environment and biological systems. According to experts' estimate, human activities and climate change are driving species to extinction at 1,000 times the natural rate. About 34,000 species of plants and more than 5,200 species of animals are on the verge of extinction worldwide, and the number and distribution of species have been undergoing dramatic changes (Geng 2013), plungeing the Earth into a global ecological crisis.

The crisis of biodiversity is a result of multiple factors. In 2010, the United Nations Environment Programme and the Secretary of the Convention on Biological Diversity (CBD) released the *Global Biodiversity Outlook 3*, indicating that the five main factors that directly cause biodiversity loss are: habitat change, overexploitation, pollution, invasive alien species and climate change (UNEP and SCBD 2010); and the rate of species extinctions caused by human activities is more than 1,000 times greater than their natural rate (Pimm et al. 2014). Among all the human activities, industrialization and urbanization are the primary culprits. Disorderly urban development has shrunk many wildlife habitats; the construction of regional infrastructure such as railways and highways has fragmented wildlife habitats, directly threatening the reproduction of wildlife species; the construction of water conservancy facilities, especially sluices and dams, has separated rivers and lakes, blocking the channels for

J. Gan, *Urban Biodiversity and Built Environment*,
https://doi.org/10.1007/978-981-16-0949-7_1

fish migration and communication between populations; the extensive use of pesticides and fertilizers in agricultural production, as well as the arbitrary discharge of industrial waste and domestic garbage, have changed the physiological characteristics and habitats of biological species, leading to the extinction of many species or a significant reduction in their populations.

Compared with other global environmental issues, biodiversity loss requires more urgent attention. According to the *Global Biodiversity Outlook 3*, "The continued loss of biodiversity has major implications for current and future human well-being" (UNEP and SCBD 2010). Whether biodiversity loss will damage the biogeochemical cycle of ecosystems, and whether the ecosystems closely related to human health are healthy, have long been topics of widespread concern in ecology (Xu and Ye 2002). "Ecosystems and Human Well-being: Biodiversity Synthesis" of the *Millennium Ecosystem Assessment* released by the United Nations Environment Programme (UNEP) released in 2005 pointed out that the continuous loss of biodiversity has worsened ecosystem services, thereby increasing the fragility of ecosystems, reducing food supply, and greatly affecting human health (UNEP 2005). The main underlying reason for this is that changes in biodiversity alter the processes of ecosystems as well as their resistance and resilience in the face of environmental changes, exerting a profound influence on ecosystem services and further affects human health and well-being through a feedback mechanism (UNEP and SCBD 2010). **It can be concluded that human activities, especially rapid urbanization, not only affect the quality of the urban ecosystem, but also has an indirect impact on the physical and mental health of urban residents.**

1.1.2 The Significance of Biodiversity in Urban Environment is Often Underestimated

Biodiversity is essential for cities, and is vital to maintaining the ecological balance and sustainable development. Cities, which are home to large populations, are commonly believed to have a poor biodiversity of flora and fauna. However, all the six elements, namely nutrients, water, sunlight, air, proper temperature and living spaces, required for the existence of living things are available in cities. Therefore, regardless of geographic location and climate, many cities harbor relatively rich diversity of species in terms of vascular plants and animal groups. Some of the cities are even located in (or close to) areas with biodiversity hotspots, while others are important transit habitats for migratory species. According to relevant research of Europe, more than 50% of the species communities of a region or even a country can be found in cities. For example, 50% of the flora in Belgium can be found in Brussels; more than half of the bird species in the surrounding landscape of Rome can be found in the city itself; and 50% of vertebrates and 65% of birds in Poland can be found in Warsaw. In the past 10 years, at least 12 new species of birds and 2 new species of mammals have been discovered in highly urbanized areas (Müller and Werner 2010). As can be seen, cities are not necessarily "deserts" with very few wildlife, and they are able to accommodate other species that coexist with humans.

Despite the numerous ecosystem services provided by urban biodiversity for the built environment, its supporting service and value tend to be underestimated. In terms of ecosystem services, in addition to their aesthetic value and cultural service, they regulate the supply and quality of water, air, and soil, and provide such services as increasing air humidity, remediating contaminated soil, improving soil fertility, and reducing noise. The water supplied to the urban areas of a city usually comes from water catchments inside and outside the territory of the city. These water catchments keep functioning properly thanks to the natural ecosystems that can store and purify water. Urban landscaping oxygenates, sequestrates carbon, absorbs solar radiation, reduces air pollution, maintains water balance, and reduces the heat island effect by adjusting the surface temperature of urban landscapes through shading and evapotranspiration. Parks and natural areas not only support the natural ecological subsystems of cities and the habitats of various species, but also provide residents with opportunities for leisure and education, encourage citizens to be in close contact with nature, and create a sense of place, which has a positive impact on human physical and mental health (Gómez-Baggethun et al. 2013). As for economic value, studies of the UK and the US also show that street trees, visible natural landscapes and waters can increase the value of properties by 5 to 18% (University College Dublin 2008).

It is true that urban biodiversity also generates certain disservices, such as decomposition and decay of wood structures caused by microorganisms, erosion of stone buildings and sculptures by bird excrement, diseases transmitted by mosquitoes, and vision blocking by trees. However, **urban biodiversity is important to maintaining the ecological security and balance of the urban system, and improving the urban settlement as it can be directly perceived and sensed by humans, and is also a key indicator to measure the urban ecological environment**.

1.1.3 The Role of Cities in Biodiversity Conservation is Gaining Increasing Attention

As the world enters a new era of rapid urbanization, cities are playing an increasingly important role in protecting biodiversity. Effective use of urban land and management of natural ecosystems can benefit both the residents and biodiversity in and around a city. For that reason, cities have become an important part of the solutions to curb global biodiversity loss. In 2006, the ICLEI–Local Governments for Sustainability Conference was held in Cape Town, South Africa. More than 300 representatives of ICLEI member cities and local authorities gathered together to establish a pilot program, the ICLEI-Local Action for Biodiversity (ICLEI-LAB). Later, in March 2007, Cities and Biodiversity: Achieving the 2010 Target, Mayors' Meeting on the Contribution of Cities to the Achievement of the 2010 Biodiversity was held in Curitiba, Brazil. The meeting launched the "Global Partnership on Cities and Biodiversity" to support local authorities in the sustainable management of their biodiversity resources, to assist cities in implementing national and international strategies on biodiversity, and to provide a platform for cities to share best practice solutions. In 2010, the partnership was further expanded and renamed as "Global Partnership on Local and Subnational Action for Biodiversity" with the goal of bringing together other levels of local and subnational (provincial and prefectural) authorities and establish the "Network of Regional Governments for Sustainable Development (nrg4SD), an international organization composed of about 50 subnational (provincial and prefectural) governments from 30 countries. This partnership has been vigorously supported and promoted by the Secretary of the Convention on Biological Diversity (SCBD), and many other city alliance organizations have joined in it, such as the World Mayor's Council on Climate Change, eco-city projects and the Urban Biosphere Network (URBIS), Urban Biodiversity & Design (URBIO), etc. (Singapore National Parks Board 2014).

The ninth meeting of the Conference of the Parties to the Convention on Biological Diversity (COP 9) was held in Bonn, Germany in May 2008, marking a major milestone since the convention was concluded in 1992-the role of cities and local authorities in protecting and strengthening global biodiversity finally won recognition. Decision on cities and biodiversity (Decision IX/28) adopted by COP 9 encourages governments to involve their cities in the implementation of the *Convention on Biological Diversity*. In addition, Decision IX/28 also provides favorable conditions for cities, local authorities and subnational governments to enable them to involve more in the planning work of local authorities concerning the *Convention on Biological Diversity*. The importance of urban biodiversity was officially confirmed by the UN and contracting parties for the first time. The "Local Action for Biodiversity" conference held in Bonn on May 26–27 of the same year was attended by mayors from 50 cities in 30 countries. The "Urban Biodiversity and Design: Implementing the Convention on Biological Diversity in Towns and Cities" conference held in Erfurt, Germany from May 21–24 invited more than 400 scientists, planners and other practitioners from 50 countries. The relationship between scientific knowledge & practice on biodiversity and urban environmental planning, design & management was discussed for the first time at the conference. The collection of papers of the same name as the conference published afterwards also points out that urban biodiversity is important because: ① it is distinctive but extremely important for the biodiversity of the entire planet; ② it reflects human culture; ③ it contributes to the quality of life in an increasing global society; ④ people have the instinct to be in close contact with other living things in nature, and urban biodiversity is the only biodiversity that most humans can experience (Müller and Werner 2010).

At the tenth meeting of the Conference of the Parties to the Convention on Biological Diversity (COP 10) held in 2010, the parties to the convention adopted Decision X/22 Plan of Action on Subnational Governments, Cities and Other Local Authorities for Biodiversity to support countries and local authorities in implementing biodiversity strategies and plans from 2011 to 2020 by offering national governments specific suggestions on how to increase the engagement of local authorities and how to integrate national strategies with local realities. The meeting also proposed to encourage

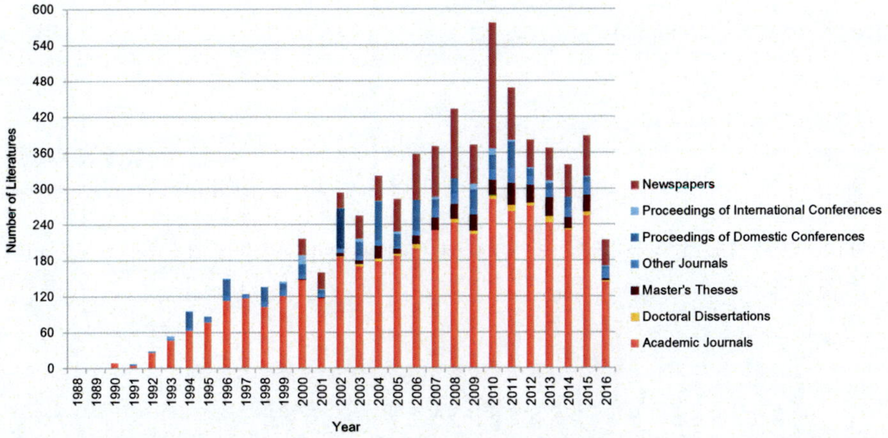

Fig. 1.1 As of August 15, 2016, literature on CNKI containing "biodiversity" in the title. *Data source* http://www.cnki.net/

the use of the City Biodiversity Index (CBI) as a monitoring tool to assist local authorities in assessing their progress in urban biodiversity conservation.

Based on a study of international academic papers about global urbanization from 2006 to 2015 by the Tongji University Library, "biodiversity," "land use," and "conservation" have become the only three that appear every year among the five urbanization research keywords in the past decade (Shen et al. 2015), which is a testament to the increasing prominence of biodiversity in urbanization research.

In China, research on urban biodiversity is also receiving increasing attention. A search on CNKI shows that as of August 15, 2016, among the scientific and technological literature (including journals, papers, master's theses, doctoral dissertations, conference papers at home and abroad, and newspapers), 6,638 contained "biodiversity" in the title, 99 contained "urban biodiversity" in the title, accounting for 1.5% of the total, and 58 were published after 2008, which shows the influence of "urban biodiversity" as an international trending topic on Chinese researchers in the field (Figs. 1.1 and 1.2).

1.1.4 The Eco-City Planning in China Lacks Research on the Relationship Between Urban Biodiversity and Built Environment

In recent years, amid China's rapid urbanization, vigorous efforts have been put in the construction of ecological civilization, and cities across the country have been striving to build themselves into an "eco-city," a "low-carbon city" or a "low-carbon eco-city." According to the *Research Report on China's Low-carbon Eco-cities* published by

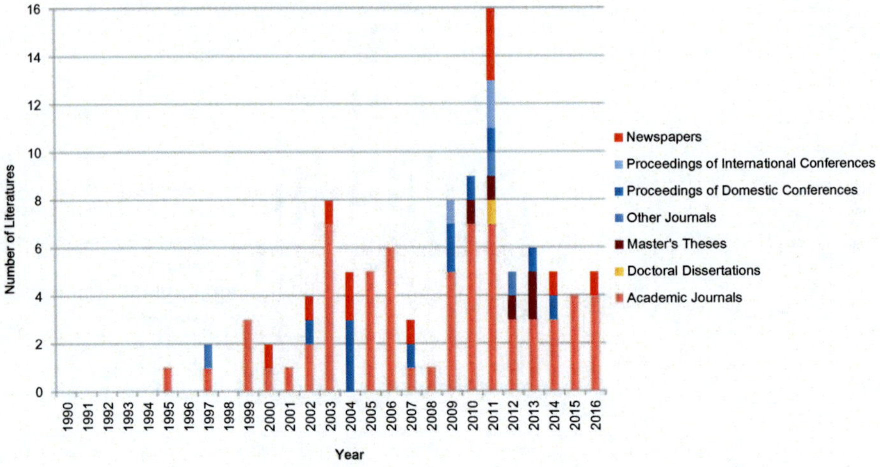

Fig. 1.2 As of August 15, 2016, literature on CNKI containing "urban biodiversity" in the title. *Data source* http://www.cnki.net/

the Chinese Society for Urban Studies, a total of 259 prefecture-level cities have proposed to develop themselves into an eco-city or a low-carbon city, accounting for more than 90% of the total number of prefecture-level cities in China. Among others, pilot projects such as the Sino-Singapore Tianjin Eco-city, the Tangshan Caofeidian Eco-city, the Shenzhen Guangming District, and the Wuxi Taihu New City have made valuable attempts to build an eco-city. **However, in terms of the investigation and diagnosis of the current ecological base, the ecological planning and design methods, or the eco-city evaluation indicator system, most of the eco-city planning research and construction mainly emphasize the creation of a green environment from the perspectives of transportation, energy, buildings, etc., while the research and practice concerning urban biodiversity and their relationship with the built environment have been given less attention.**

Taking the eco-city indicator system as an example, the *Construction Indicators for Ecological Counties, Ecological Cities, and Ecological Provinces* released by the Ministry of Environmental Protection includes 4 indicators for economic development, 11 indicators for environmental protection, and 7 indicators for social progress, totaling 22 indicators in three categories, but none of the indicators involve "biodiversity"; the *Index for Low-carbon Eco-cities* prepared by the Chinese Society for Urban Studies contains 7 indicators for resource conservation, 9 indicators for environmental friendliness, 10 indicators for social harmony, and 4 indicators for economic sustainability, among which only the "comprehensive species indicator" and the "local species indicator" involve "biodiversity." The evaluation indicator systems of the Sino-Singapore Tianjin Eco-city, the Tangshan Caofeidian Eco-city, the Shenzhen Guangming District, the Wuxi Taihu New City, the Shanghai Chongming Eco-Island, etc. formed on this basis include a few indicators in the biodiversity dimension, such as the "local plant indicator," "species diversity of new green space," "tree coverage of new green space," "number of waterfowl species accounting for more than 1% of the global population of species" and so on. But on the whole, **these indicators focus on plants and overlook animals. Some indicators are about specific species and lacks spatial guidance. As a result, it is difficult to integrate them into the urban planning control and management system, and thus impossible for them to truly play a role in evaluation, implementation, and management**.

Europe and some other countries place great emphasis on biodiversity conservation in urban planning. In all stages of urban planning, biodiversity is listed as an important content, and various planning policies are formulated based on large-scale surveys. For example, *A Biodiversity Guide for the Planning and Development Sectors in South East England (2002)* of the UK, the *Biodiversity Planning Guidelines for the Perth Metropolitan Region (2004)* developed by the Western Australian Local Government Association, the *National Biodiversity Strategy and Action Plan (*2009*)* developed by the National Parks Board (NParks) of Singapore, the *Berlin Strategy on Biological Diversity (*2012*)* issued by the German Senate Department for Urban Development and the Environment, (Shen 2004; Zhao 2013; Singapore National Parks Board 2009; Senatsverwaltung für Stadtentwicklung und Umwelt 2012) and others all emphasize the important relationship between urban planning and biodiversity conservation, aiming to monitor the impact of land use changes on biodiversity in a long run, help planner and designers to combine biodiversity data with land use data, and propose specific approaches and methods to protect biodiversity in the planning system.

Since the former Ministry of Construction promulgated the *Notice on Strengthening Urban Biodiversity Protection* in 2002, the important role of urban planning in biodiversity conservation has been gaining a growing recognition in China. But in general, the research and practice concerning biodiversity in the field of urban planning in China is still in its infancy. At present, some cities have prepared urban biodiversity conservation plans on the macroscale. However, due to lack of relevant preparation norms and insufficient awareness and capability of competent departments, these plans have not been well implemented. On the meso-micro scale, to meet the assessment requirements for the environmental impact of planned projects and projects under construction, or thanks to the increasing awareness of biodiversity conservation, urban biodiversity surveys have been implemented in the construction of major projects in some cities. For example, in the initial stage of planning and construction of World Expo 2010 Shanghai, a detailed survey was carried out by the East China Normal University and the Shanghai Science and Technology Museum on the status of singing insects and butterflies in the pan-Expo Park area, bird diversity and influencing factors in the Expo Park and its surrounding areas, vegetation and plant resources in the Expo Park, and plant diseases and insect pests & regional rodents in the Expo Park, and an information database was established (Wang et al. 2008); The Tianjin Natural History Museum and the Shanghai Science and Technology Museum conducted surveys on the biodiversity in the bases of the Sino-Singapore Tianjin Eco-city and the Shanghai New Jiangwan Town when the projects just started (Guo and Wang 2008; Jin et al. 2005), etc. A mass of local biodiversity data has been accumulated through those surveys. However, these surveys are disconnected from the current control system of urban planning and design, and do not take into consideration a key feature of urban planning, namely the spatial allocation of land resources, while the planners fail to understand the relationship between biodiversity data and spatial layout. So, a large number of biodiversity surveys and studies on the biological zoning completed in the early stage cannot be brought into full play during the planning and implementation.

In fact, the concept of "eco-city" was first mentioned by the UNESCO in the "Man and the Biosphere Programme (MAB)" in 1971. This programme emphasizes the harmonious coexistence of man and the natural environment in the entire biosphere of the Earth from the very beginning. In the critical period when China's urbanization rate exceeded 50%, the report of the 18th CPC National Congress clearly put forward an ecological civilization concept, stressing that "We, as human beings, must respect nature, follow its laws, and protect it." The 3rd Plenary Session of the 18th CPC Central Committee proposed to "put systematic and complete systems for improving the ecosystem in place," and put forward clearer and higher-level requirements of ecological transformation for urban planning and construction in the new era. The *Cities and Biodiversity Outlook: Action and Policy* jointly released by the UN-Habitat and the ICLEI—Local Governments for Sustainability in 2012 pointed out that urbanization is both a challenge and an opportunity for biodiversity and ecosystem services (Convention on Biological Diversity et al. 2012), providing references and guidance for how to protect natural biological resources in the process of urbanization and urban development.

In conclusion, maintaining and improving biodiversity in a high-density urban built environment is one of the key challenges facing sustainable urban development. The role of biodiversity in urban planning, especially in the research on eco-city planning and design, entails further exploration, and sheds light on future urban ecological transformation and development.

1.2 Current Researches of Planning and Design for Urban Biodiversity with a China Focus

1.2.1 Planning for Unban Biodiversity on the Macroscale

1. **Ecological network planning**

On the macroscale, the key of systematic planning for biodiversity and habitats protection lies in the construction of an ecological network system. *The Pan-European Biological and Landscape Diversity Strategy* proposes to establish a trans-European ecological network system for biodiversity protection, which involves different spatial scales, and urban areas are an important part of it (European

Commission 1992). Taking the United Kingdom as an example, the "Sites of Special Scientific Interest" (SSSIs) at the national level are included in the legal protection scope as the best wildlife sites. Among them, SSSIs with special quality are designated as national nature reserves. At the local and urban levels, important nature reserves or wildlife sites are ranked according to their importance in the statutory spatial regulation developed and planned by localities, so as to promote education about nature and raise public awareness (Department for Food and Rural Affairs 2011). While in the Netherlands, a country with a large population and limited land, the goal is to incorporate a large number of small habitats into the protection network of the overall system. The National Ecological Network, which kicked off in 1990, aims to integrate fragmented habitats through the "reorganization of fragments" of major nature reserves, natural restoration areas, and ecological corridors (Liu 2009).

In China, the research and practice concerning ecological network planning mainly include: Comprehensive Greenspace Planning Based on Landscape Ecology Principles in Compact Nanjing City, China (Jim and Chen 2003); Comprehensive Concept Planning of Urban Greening Based on Ecological Principles: A Case Study in Beijing, China (Li et al. 2004); Research on Urban Ecological Network: Taking Xiamen as an Example (Wang 2005); Ecological Planning: Methods and Implementation Mechanism in Megacities (Zhan and Li 2011); etc. Unlike Europe where the maintenance of biodiversity and the protection of wildlife habitats are given top priority, the ecological network planning in China focus on the delineation of urban growth boundaries (UGB) and the provision of sufficient ecological space for basic ecosystem services. For this reason, the ecological network planning in China is slightly inadequate in maintaining biodiversity, and at the same time lacks the connection with and the extension of regional ecological networks.

2. **Conservation Planning for Urban biodiversity**

Conservation Planning for Urban biodiversity is another major model for protecting wildlife and their habitats on the macroscale. Berlin, Germany, introduced the "Biotope Area Factor" strategy in 1994 to build urban green infrastructure while preserving housing density. The planned habitat network covers the entire city, and the central city retains or increases areas that can reach nature while maintaining housing density. The transition area provides connected habitats that can serve a wider area, while the urban fringe ensures that large habitats are embedded in the urban area in the shape of "fingers" (Berlin Department of Urban Development 1995). Sydney, Australia has proposed to introduce wild flora and fauna into the city to protect and enrich biodiversity, and preserved and restored large areas of wetlands, swamps and jungles in the Olympic Park. The Netherlands has enabled its cities to take on a new look through plant planning. Japan, to protect urban biodiversity, has "introduced nature into cities," that is, introducing natural community structure mechanisms into cities or establishing artificial communities with similar structures (Zhao and Liu 2009).

Currently, Conservation Planning for Urban biodiversity in China is mainly about the greenspace system planning in the master planning stage of cities, and focuses on the planning for the conservation of plant diversity. Related research mainly includes: The Countermeasures of Biodiversity Conservation Project in Quanzhou City (Huang 2002); Studies on Plant Biodiversity Conservation and Construction Planning in Green Space of Quzhou City (Wang et al. 2005); Urban Plant Diversity and Conservation Planning in Sanming County (Zhou 2008); Planning of Biodiversity Conservation in Jinshan District (Yang and Cao 2005); Study on Urban Biodiversity Conservation Planning of Wuxi City (Cui 2011); Research on Conservation Planning of Urban Biodiversity in Guangzhou City (Cao 2013), etc. In addition, cities such as Shijiazhuang, Kunming, Haimen, Wuzhou, Fuzhou, Huangshan, Anshun, Lishui, etc. have also prepared their own urban biodiversity conservation plans at the urban master planning level in recent years.

3. **Local Biodiversity Action Plan**

The Local Biodiversity Action Plan (LBAP) is a pilot project launched by the ICLEI—Local Governments for Sustainability in 2006. As a local action guide for local authorities under the framework of national biodiversity planning goals, LAP has gained the attention of many cities in other countries.

In the UK, LBAP is jointly prepared by local authorities and stakeholders to determine specific habitats and species that need preferential protection. According to the protection priorities determined by the national plan, the protection of important habitats and critically endangered species takes precedence, whereas the general habitats and native species takes second place (Environment Agency et al. 2009).

In Ireland, the County Development Plan highlights designated sites, protected species, and ecologically sensitive landscapes. It puts forward policies and goals for conserving and strengthening biodiversity in wild landscapes, to protect rivers, small woodlands and species-rich grasslands and other broader substrates. LBAP is at the core of the County Development Plan and its main task is to conduct a general survey of habitats and assessment of biodiversity at the county level to determine the habitats, species and sites of local importance, paving the way for the subsequent delineation of sensitive areas. As a legal plan, LBAP incorporates biodiversity considerations into the integrated framework of site development (University College Dublin 2008).

The LBAP of Singapore, a high-density Asian city, emphasizes the transition from a "Garden City" to a "City in the Garden." Singapore aims to build itself into "a Haven for Biodiversity," views biodiversity resources as national natural heritage, and adopts integrated nature conservation methods to build an ecospace network in which man and nature coexist harmoniously. It also takes biodiversity and ecosystem into consideration, and incorporates them into the national planning (Singapore National Parks Board 2009).

1.2.2 Planning for Unban Biodiversity on the Mesoscale

Planning for unban biodiversity on the mesoscale mainly aims to control and guide biodiversity conservation on the scale of urban communities, which is generally implemented as a specific strategy for macro-level planning. The Town and Country Planning Association of the United Kingdom issued the *Biodiversity by Design: A Guide to Sustainable Communities* in 2004, proposing to protect and enhance the biodiversity of communities by establishing new green infrastructure, green grids and community forests, parks and natural spaces, greenway linkages and street trees, and putting forward different development density recommendations for growth areas of sub-region and new communities as well as housing market renewal pathfinder areas (Town and Country Planning Association 2004).

The Viikki District of Helsinki, Finland, lays stress on the connection with nature on the community scale. It has planned 1,700 residential units close to nature and adjacent to a 250 hm^2 wetland of bird nature reserve, built a 34 hm^2 community park, and introduced natural areas in the shape of "fingers" into each residential group through ecological planting corridors. Furthermore, it has promoted the integration of nature education and community through the construction of children's ecological parks, garden centers and allotment gardens (City of Helsinki 1999).

Malmö, Sweden, explicitly incorporates requirements for the control and guidance of biodiversity in its development control. For example, in the planning of its new district of Bo01, the concepts of "habitat mosaics" and "green space factors" were proposed, and each property developer was required to adopt measures to enhance biodiversity and manage rainwater. Each developer had to choose 10 out of 35 Green Points which include: at least 50 species of native herbs in the courtyard; all walls are covered with climbing plants; all roofs are green roofs; each apartment has at least one bird's box; there are food for birds in the yard all year round in the courtyard; there is installation facilitating swallows' nesting on the building façades; there are bat boxes and habitat for specified insects in the courtyard, etc. (Town and Country Planning Association 2004).

1.2.3 Planning and Design for Biodiversity on the Microscale

On the microscale, Connery (2009) proposes a design method of integrating regional ecological networks at the block and site scales; AECOM (2013) proposes the Landscape Biodiversity Index, which includes dimensions such as habitat priority, diversity, quality, area, form and scale, connectivity, and ecosystem type and pattern; Wang and Song (2014) propose a biodiversity design framework for urban parks based on the classification system of habitat units in *Biodiversity Design for Urban Parks*.

Some other studies focus on the restoration and creation of habitats of target species with certain ornamental value such as birds and fishes. The types of spaces involved mainly include park, wetland and open waterfront space. The studies include: Research on Design of City Lake Park Based on Fish Habitat Repair (Pan 2013); Research on Design of Urban Waterfront Based on Bird Habitat Repair (Liu 2015); and Lin (2001)'s design method based on the biodiversity of different species such as birds, fishes, and fireflies from the perspective of habitat reconstruction proposed in his *Urban and Rural Ecology*.

There are also some studies that propose design strategies in terms of vegetation configuration and three-dimensional greening, including: the theory and method of community ecological design proposed in *Community Ecological Design* (Wang et al. 2009); the biophilic design strategies in high-density cities such as biodiversity roofs, ecologically responsible facades, and biophilic sky gardens proposed in *Research on Biophilic Design Mode in High-rise and High-density City* (Ye and Yu 2016); the concept of "integrating micro-nature in design," that is, incorporating micro nature protection in the architectural design for site layout, appropriately and partially naturalizing the green space in the built environment, setting up close-to-nature green spaces in the built environment, and carrying out micro-natural restoration with plant communities to improve urban biodiversity and ecological stability, which is proposed in *Design With Micro-nature* (Wu et al. 2016).

In practice, many cities formulate design guidelines or rules to support biodiversity design on the microscale. For example, Islington District in the Greater London proposed in the *Sustainable design and green planning good practice guide 4: biodiversity in the built environment*: Even where there are no significant habitats or protected species present on a site, biodiversity is still an important consideration. All new development should contribute to the enhancement of biodiversity and create habitat wherever possible, from green roofs, green walls, and small measures such as installation of a nest box, to more comprehensive actions such as design of a fully integrated ecological landscape and Sustainable Drainage Systems (SUDs). It lays emphasis the design and improvement of small-scale biodiversity, such as the use of community gardens, pocket gardens and other "doorstep" spaces to create potential for better habitat modules (Islington County Council 2012). Gunnell et al. (2013) propose guidelines for customizing recreational habitats and nesting environments for urban wildlife in new buildings and the reconstruction of existing buildings, and effectively preventing light pollution and vehicle collisions. San Francisco, New York, and Minnesota in the United States have all launched bird-friendly city site planning, landscape design and architectural design guidelines to ensure bird safety (New York City Audubon Society 2015; Audubon Minnesota 2010; San Francisco Planning Department 2011). In a report provided to the German Federal Agency for Nature Conservation, Werner et al. (2016) analyze practical cases of integrating the habitats of wild fauna such as swallows, pigeons, and bats in the renovation and reconstruction of buildings.

1.2.4 *Deficiencies in Current Planning and Design for Unban Biodiversity in China*

1. **Planning perspective–stressing conservation but overlooking enhancement**

When it comes to technical terms, China's planning strategies for unban biodiversity unexceptionally set the ultimate goal of "conserving," whereas many Western countries place equal emphasis on "conserving" and "enhancing" biodiversity, which mirror the different attitudes towards the use of urban biodiversity and habitat spaces. Albeit the increasing attention on planning for unban biodiversity in China, the research and practice concerning biodiversity planning strategies still play a comparatively insignificant role in the entire planning system as many pressing issues regarding urban society and disaster prevention need to be resolved first. Moreover, after 40 years of high-speed, high-density, and high-intensity urbanization, much of the ecological spaces supporting biodiversity have been occupied in China. Therefore, the primary role of biodiversity planning is still to "defend the ecological red line." Western countries, which are already in steady urbanization, are able to meet the needs of urban residents to get close to other living creatures, and therefore strive to enhance biodiversity and its accompanying ecosystem services while conserving biodiversity.

2. **Planning area–focusing on parts but overlooking the whole picture**

Unlike the research and practice in China that pay more attention to nature reserves, large-scale parks and green spaces, and habitats of specific species, Western countries attach more importance to the potential of biodiversity for full coverage of all urban spaces, and this is especially true in the study of planning strategies. The "Go Wild" of the UK, "Going Native" of the US, "Introducing Nature into Cities" of Japan, and a "City in the Garden" of Singapore all embody the recognition of and respect for the value of urban biology, urban nature, and wilderness in cities. Numerous planning and design guidelines for the conservation and enhancement of biodiversity have been issued by planning departments and agencies at all levels, design companies, and NGOs, covering multiple scales such as regions, cities, communities, and buildings and enabling planners, developers, and proprietors to jointly build an urban environment in which humans and other creatures coexist through their own efforts.

3. **Planning scale–focusing on both ends but overlooking the connection**

In regard to planning scale, both researches and practices pay more attention to the macro-scale and the micro-scale planning, and less to meso-scale research and practice. This is particularly true in the research in China. In fact, the mesoscale is an indispensable and important link in the entire planning system, as it undertakes and implements the macro-scale goals and reserves space for the microscale. In particular, those cities in China with high-density buildings, under the guidance of the strategy of building a country with ecological civilization, increasing attention has been paid in recent years to the delineation of basic ecological spaces and ecological networks on the macroscale, laying a solid foundation for biodiversity protection in the new round of urban development. The mesoscale, as the frontier directly facing development and construction, will be the most vulnerable link in the biological habitat protection without more attention to biodiversity, which will inevitably make the habitat creation more fragmented and lead to greater impact on the overall biodiversity and ecosystem services.

4. **Planning object–focusing on plants but overlooking animals**

In the planning and practice concerning unban biodiversity, especially in the conservation planning and practice at the master plan level, research in China pays more attention to plants and less to urban wild animals. The eco-city indicator system also focuses on biotope indicators such as the "local plant indicator" that emphasizes plant diversity. However, wild animals are an indispensable part of a city, bringing endless vitality and giving people more opportunities to get close to nature and relax after work. More importantly, wild animals are at a higher level among the urban biological trophic groups. Compared with plants featuring a higher degree of artificialization, the spatial behaviors of wildlife animals in a city are a result of their own adaptation to the urban ecosystem, serving as a better indicator of the urban ecological environment. For example, frogs are indicator species for the water quality of a city's urban water system, dragonflies are indicator species for its wetland habitats, and birds are indicator species for the comprehensive quality of its urban ecosystem. Therefore, urban wildlife should also be taken into account in the planning for unban biodiversity.

1.3 Key Points of the Book

Based on the analysis of the current research background, as well as the author's analysis and summary of the available research, the author finds that there are still many problems waiting to be solved in the field of urban biodiversity and how to integrate it into spatial planning and design. For example, with high-density urbanization becoming a development option for most large cities in China, how should we balance the relatively more "compact" urban human settlements and wildlife habitats? Urbanization is both a challenge and an opportunity for biodiversity. How will the changes in the spatial elements of the built environment brought about by urban development affect urban biodiversity? Are there any special habitats in cities that are different from those in natural conditions? Is there a threshold for biological habitat space in the existing layout and pattern of urban planning and land use?Is it possible to optimize and improve the biodiversity of high-density cities through urban spatial planning and design methods? Facing such an enormous and complex system, we need to answer a basic question before coming up with answers to these questions: in the development of China's relatively high-density urban spatial pattern, are the elements of the urban built environment at different scales related to biodiversity? If the answer is yes, in which dimensions are they related? And what is the underlying mechanism?

To find answers to these questions, this book systematically analyzes the relationship between urban built environment and urban biodiversity at different scales through theoretical and empirical research, extracts key indicators of control guidance for planning and design based on spatial structure variables, and proposes supplementary planning and design strategies. The author wrote this book with the hope to lay the theoretical groundwork for the preparation, implementation and management of urban ecological planning, offer technical support for improving urban ecosystem services in practice and for ensuring the harmonious coexistence of man and nature in the sustainable urban development, and provide a basis for urban spatial planning and design to balance the contradiction between the development of "high-density and compact" urban areas and the conservation and enhancement of biodiversity in the process of China's urbanization, so as to promote the "harmony between man and nature" in sustainable urban development.

References

AECOM (2013) Landscape biodiversity planning & design system (Technical Report)

Audubon Minnesota (2010) Bird-safe building guidelines. http://mn.audubon.org/sites/default/files/05-05-10_bird-safe-building-guidelines.pdf. Accessed 1 Sept 2016

Berlin Department of Urban Development (1995) Valuable areas for flora and fauna. www.stadtentwicklung.berlin.de/umwelt/umweltatlas/ei503.htm. Accessed 1 Sept 2016

Cao XX (2013) Research on conservation planning of Urban biodiversity in Guangzhou City. Guangzhou: Master thesis of Zhongkai College of Agricultural Engineering (in Chinese)

City of Helsinki (1999) Viikk-a university district and science park for the 2000's. City Planning Department

Connery K (2009) Biodiversity and urban design: seeking an integrated solution. J Green Build 2:23–38

Convention on Biological Diversity et al (2012) Cities and biodiversity outlook: action and policy. https://www.cbd.int/doc/health/cbo-action-policy-en.pdf. Accessed 1 Sept 2016

Cui RZ (2011) Study on Urban Biodiversity conservation planning of Wuxi City. Nanjing: Master thesis of Nanjing Agricultural University (in Chinese)

Department for food and Rural affairs (Defra) (2011) UK biodiversity 2020: a strategy for England's wildlife and ecosystem services. https://www.cbd.int/doc/world/gb/gb-nbsap-v3-p1-en.pdf. Accessed 1 Sept 2016

Environment Agency, Brighton & Hove City Council, the Bat Conservation Trust, et al (2009) Biodiversity positive: eco-towns biodiversity worksheet. Town and Country Planning Association, London. https://www.tcpa.org.uk/Handlers/Download.ashx?IDMF=2e0ffaf8-24b1-45fe-a02f-505a06d72ff2. Accessed 1 Sept 2016

European Commission (1992) Council Directive 92/43/ E E C of 21 May 1992 on the Conservation of Natural Habitats and of Wild Fauna and Flora. http://ec.europa.eu/environment/nature/legislation/habitatsdirective/index_en.htm. Accessed 1 Sept 2016

Geng GB (2013) Life is so wonderful. Green China 8:10–21 (in Chinese)

Gómez-Baggethun E, Gren Å, Barton DN et al (2013) Urban ecosystem services. In: Fragkias M, Goodness J, Güneralp B et al (eds) Urbanization, biodiversity and ecosystem services: challenges and opportunities. A global assessment. Springer, Netherlands, Dordrecht, pp 175–251

Gunnell K, Murphy B, Williams C (2013) Designing for biodiversity: a technical guide for new and existing buildings, 2nd edn. RIBA Publishing, London

Guo Q, Wang QL (2008) Investigation of biological resources in Sino-Singapore Tianjin Eco-city. J Anhui Agric 36(33):14705–14706 (in Chinese)

Huang G (2002) The countermeasuesr of biodiversity conservation project in Quanzhou City. Chin J Eco-Agric 10(4):101–102 (in Chinese)

Islington County Council (2012) Sustainable design and green planning good practice guide 4: biodiversity in the built environment

Jim CY, Chen SS (2003) Comprehensive greenspace planning based on landscape ecology principles in compact Nanjing city, China. Landsc Urban Plan 65(3):95–116

Jin XB, Zhou BC, Qin XK, et al (2005) Biodiversity of Shanghai Jiangwan Airport. In: MA KP (ed) China's biodiversity conservation and research progress. Meteorological Press, Beijing, pp 394–429 (in Chinese)

Li F, Wang RS, Paulussen J et al (2004) Comprehensive concept planning of urban greening based on ecological principles: a case study in Beijing, China. Landsc Urban Plan 72(4):325–336

Lin XD (2001) Urban and rural ecology, 2nd edn. Chan's Arch-Publishing Co., Taipei (in Chinese)

Liu HL (2009) Connection and cooperation: the ecological network planning based on experiences in Europe and the Netherlands. Chin Landsc Archit 25(9):31–35 (in Chinese)

Liu JN (2015) Research on design of Urban waterfront based on bird habitat repair. Master thesis of Zhejiang Agriculture and Forestry University (in Chinese)

Müller N, Werner P (2010) Urban biodiversity and the case for implementing the convention on biological diversity in towns and cities. In: Müller N, Werner P, John G et al (eds) Urban biodiversity and design. Wiley-Blackwell, Oxford, pp 1–34

New York City Audubon Society (2015) Bird-safe building guidelines. http://www.nycaudubon.org/images/pdf/Bird-Friendly_Building_Design_2015.pdf. Accessed 1 Sept 2016

Pan CC (2013) Research on design of city lake park based on fish habitat repair. Master thesis of Zhejiang Agriculture and Forestry University (in Chinese)

Pimm SL, Jenkins CN, Abell R (2014) The biodiversity of species and their rates of extinction, distribution, and protection. Science 344(6187):1246752

San Francisco Planning Department (2011) Standards for bird-safe buildings. https://sfplanning.org/sites/default/files/resources/2019-09/Design%20Guide%20Standards%20for%20Bird%20Safe%20Bldgs_Final.pdf. Accessed 1 Sept 2016

Senatsverwaltung für Stadtentwicklung und Umwelt (2012) Berliner strategie zur biologischen vielfalt. https://www.berlin.de/senuvk/natur_gruen/biologische_vielfalt/de/berliner_strategie/index.shtml. Accessed 1 Sept 2016

Shen JH, et al (2015) International research situation in the field of urbanization. Research Report of seed fund by China Intelligent Urbanization Co-creation Center for High Density Region (in Chinese)

Shen QJ (2004) Land use planning and biodiversity-a review on "a biodiversity guide for the planning and development sectors in South East England". Urban Plan Forum 2:85–89 (in Chinese)

Singapore National Parks Board (2009) Conserving our biodiversity: Singapore national biodiversity strategy and action plan. https://www.nparks.gov.sg/biodiversity/our-national-plan-for-conservation/national-biodiversity-strategy-and-action-plan. Accessed 1 Sept 2016

Singapore National Parks Board (2014) User's manual on the Singapore Index on cities' biodiversity. https://www.nparks.gov.sg/-/media/nparks-real-content/biodi-versity/singapore-index/users-man-ual-on-the-singapore-index-on-cit-ies-biodiversity.pdf. Accessed 1 Sept 2016

Town and Country Planning Association (2004) Biodiversity by design: A guide for sustainable communities. http://urbed.coop/projects/biodiversity-design. Accessed 1 Sept 2016

United Nation Environment Program (2005) A summary of the millennium ecosystem assessment: biodiversity synthesis. UNEP. https://www.millenniumassessment.org/documents/document.354.aspx.pdf. Accessed 1 Sept 2016

United Nation Environment Program, Secretary of the Convention on Biological Diversity (2010). Global biodiversity outlook 3. https://www.cbd.int/gbo3/. Accessed 1 Sept 2016

University College Dublin (2008) Green city guidelines: advice for the protection and enhancement of biodiversity in medium to high-density urban development. UCD Urban Institute Ireland, Dublin. http://www.bioone.org/doi/full/10.1641/B571005. Accessed 1 Sept 2016

Wang HZ (2005) Research on urban ecological network: Taking Xiamen as an example. Shanghai: Master Thesis of East China Normal University, 2005 (in Chinese)

Wang M, Song Y (2014) Biodiversity design for Urban parks. Lands Archit 1:47–52 (in Chinese)

Wang XD, Ma J, Zhang WR (2005) Studies on plant biodiversity conservation and construction planning in green space of Quzhou City. J Zhejiang Univ (Agric Life Sci.) 31(4):439–444 (in Chinese)

Wang XM, Wang TH, Liu YN (2008) Research on ecological elements of urban area and construction of information database–Taking Shanghai World Expo area as an example. China Science Publishing & Media Ltd., Beijing (in Chinese)

Wang YC, Han LY, Wang CP (2009) Community ecological design. China Construction Industry Press, Beijing (in Chinese)

Werner P, Groklos M, Eppler G, et al (2016) Schutz gebäudebewohnender tierarten vor dem hintergrund energetischer gebäudesanierung. Hintergründe, Argumente, Positionen. in Städten und Gemeinden. Technical Report for BFN. https://www.bfn.de/fileadmin/BfN/planung/siedlung/Dokumente/Gebaeudebruetende_Tierarten_2016_-_Positionspapier.pdf. Accessed 1 Sept 2016

Wu ZW, Shan HN, Wang YH (2016) Design with micro-nature. Huazhong Archit 5:31–35 (in Chinese)

Xu KY, Ye WH (2002) Ecosystem health and biodiversity. Ecol Sci 21(3):279–283 (in Chinese)

Yang G, Cao DJ (2005) My view on the planning of biodiversity conservation in Jinshan District. Shanghai Agric Sci Technol 5:10–11 (in Chinese)

Ye SW, Yu WJ (2016) Research on biophilic design mode in high-rise and high-density city. Hous Ind 5:26–30 (in Chinese)

Zhan YZ, Li Y (2011) Ecological planning: methods and implementation mechanism in megacities. Urban Plan Forum 2:49–57 (in Chinese)

Zhao CJ (2013) Biodiversity conservation oriented urban green spaces plan and design. In: Chinese society for urban studies(ed) Proceedings of 2013 Conference on Urban Development and Planning, Zhuhai, China (in Chinese)

Zhao MY, Liu ZL (2009) Discussion on the development status, planning approaches and methods of urban biodiversity conservation planning at home and abroad. China Popul Resour Environ 9:662–666 (in Chinese)

Zhou HH (2008) Urban Plant diversity and conservation planning in Sanming County. Fuzhou: Master thesis of Fujian Agriculture and Forestry University (in Chinese)

Chapter 2
Conceptual Framework

2.1 Concept, Type and Function of Multiple Overlapping Habitats in Cities

2.1.1 Concept of Urban Multiple Habitats

A habitat refers to a space where organisms live and breed as individuals or in populations, as well as the environmental conditions, resources, and the complicated relationships between the organisms and between the biotic and abiotic environments in that space (Whittaker et al. 1973). Cities, with their unique climate, soil, hydrology, and light, create special urban habitats that are different from the primitive environment, and produce in them specific flora and fauna (Wittig 1998). Compared with those large scenic spots, nature reserves, and national parks located in suburban areas or even farther away, urban habitats play a more immediate and important role in strengthening urban characteristics in a locality, optimizing urban ecological environment, and improving the life quality of urban residents (Han and Li 2009). Therefore, one of the major means to realize urban ecological development and ecosystem optimization is to incorporate urban habitats into urban planning and development at all levels for overall protection and sustainable use.

Although cities are dominated by humans, they do not stand in absolute opposition to nature. Copious studies have proved the existence and evolution of nature and its multiple life forms in cities. The superimposition of human activities and nature can generate infinite parts, namely the "multiple natures" (Li 2016). Based on the characteristics of urban flora and fauna, Kowarik (1992) proposed four kinds of urban nature: primary nature (urban forests and wetlands) where human ecological footprints can be ignored, rural agricultural nature, urban horticultural nature, and special nature in urban industrial areas. Urban habitats include not only the first and second types of natural ecosystems composed of flora and fauna spontaneously growing in cities, but also the third and fourth types of nature affected by artificial modification (Breuste et al. 2016).

J. Gan, *Urban Biodiversity and Built Environment*,
https://doi.org/10.1007/978-981-16-0949-7_2

The green space system and open space system in cities are of equal value to human well-being and wildlife (Pickett et al. 2001). Organisms are the "aborigines" of cities. Spaces are available in cities to meet the basic needs of urban organisms to exist and reproduce and accommodate their ecological niches. They exist not only in the singular remnant primary nature, but also in other overlapping natures disturbed by human beings to varying degrees.

In view of this, the author sees urban and rural spaces as human-nature overlapping resources, **examines the potential habitats from the perspective of "multidimensional protection," and proposes the concept of "multiple habitats," which refers to the multiple overlapping of spaces for biotic habitats and human settlements in the urban built environment on different dimensions and in different forms, indicating that biotic habitats are not restricted to primary natural spaces such as reserves and other major habitats, and also include various types of land in urban and rural environments**. The idea of "multiple habitats" argues that instead of merely being "prey," urban organisms "share" urban spaces with humans, and that urban spaces under full coverage (even 3D spaces) are potential substrates that can be used by organisms for their habitation, migration, and breeding, and to build urban biotic habitat networks with different sizes, dimensions, and functions.

2.1.2 Types of Overlapping Urban Multiple Habitats

Multiple habitats in cities, whether they be vegetated surfaces or hard grounds, vertical building walls or off-ground roof surfaces, can provide organisms with food and water, protect them from bad weather and predators, and help them attract mates and reproduce. Several studies have proposed different types of habitats which are applicable to different dimensions and different urban characteristics. For instance, based on land use classification, Gilbert (1991) divides the UK's habitats into urban commons, industrial areas, railways, roads, city centers, city parks, allotments and leisure gardens, cemeteries, gardens, rivers–canals–ponds–lakes–reservoirs and water mains, and woodland; Köstler et al. (2003) divide, based on natural conditions and land use, Berlin's habitats(biotopes) into 12 categories and 91 subcategories, including flowing waters (Fließgewässer), still waters (Standgewässer), anthropogenic crude soil sites and wild grasslands (Anthropogene Rohbodenstandorte und Ruderalfluren), moors and swamps (Moore und Sümpfe), greenland, meadow and lawns (Grünland, Staudenfluren und Rasengesellschaften), dwarf shrubs (Zwergstrauchheiden), Vegetables, rows of tree lines and groups of trees (Gebüsche, Baumreihen und Baumgruppen), forests and woodlands (Wälder und Forsten), cultivated lands (Äcker), green and open spaces (Grün- und Freiflächen), special biotopes (Sonderbiotope), built areas, transportation facilities and special areas (Bebaute Gebiete, Verkehrsanlagen und Sonderflächen); the Centre for Urban Greenery and Ecology (CUGE) in Singapore has classified the country's key landscape habitats into forests, freshwaters, coastal zones, grasslands, and urban novel habitats (Tertilt

2010); the UK National Ecosystem Assessment (UK-NEA) divides the urban habitats within the country into natural and semi-natural greenspaces, street trees, public parks and formal gardens, domestic gardens, green corridors, outdoor sports facilities, recreational areas and amenity greenspace, allotment, community gardens and urban farms, cemeteries, churchyards and burial grounds, previous developed land (brownfields), water, etc. (Bateman et al. 2011); Müller et al. (2012) argue that habitats found in the human settlements include remnant vegetation vegetation (e.g., remnant habitats of native plant communities, rock faces), agricultural landscapes (e.g., meadows and arable land), urban-industrial landscapes (e.g., wastelands and vacant lots, residential areas, industrial parks, railway areas, and brownfields), ornamental gardens and landscapes (e.g., formal parks and gardens, small gardens and green spaces); Douglas and James (2014) divide habitats into encapsulated countryside, managed green spaces, naturally regenerating habitats, and urban wetlands according to their sources.

As this book mainly discusses the relationship between the urban built environment and biodiversity, the nature of urban land use is the first thing to be considered in the classification of multiple habitats in a city. The *Standard of Urban Land Classification and Planned Development Land* (GB 50137-2011) divides urban and rural land into development land and non-development land, while the *Technical Guidelines for Controlled Detailed Planning of Shanghai* (2016 Revision) divides Shanghai's urban and rural planning land into urban and rural development land, agricultural land, waters, and unused land. Therefore, it is necessary to distinguish development land from non-development land in the classification.

Furthermore, urban multiple overlapping habitats are affected by the development activities of humans to various degrees, showing various medium characteristics. Therefore, the degree of interference of artificial environment on the primitive nature environment, i.e., the degree of human-nature overlapping, should also be taken into account in the classification.

According to the above classification bases, this book proposes to divide the multiple biotic habitats across the whole urban land of a city into near-natural habitats on the non-development land, habitats with natural-succession on the abandoned land, semi-natural habitats in the urban parks and public green spaces, semi-artificial habitats in the attached green spaces, and artificial habitats on the sealed surfaces (Fig. 2.1).

Fig. 2.1 Types of human—nature overlapping multiple habitats

Fig. 2.2 Near-natural habitats on the non-development land (Left: Shanghai Dongtan Wetland Park; Right: Water forest in Shanghai Qingxi Country Park)

(1) Near-natural habitats on the non-development land

They refer to the highly fragmented wildlife habitats that remain in the built and marginal areas of a city during long-term urban development, and the habitats that evolve on less disturbed artificial substrates. Thanks to long-term evolution, these habitats have become adapted to the local natural and artificial conditions, and co-exist in harmony with the urban environment which they are part of. Typically, they mainly include agricultural and forestry habitats on non-development land, such as woodland, cultivated land, garden land, and grassland, as well as aqua and near-aqua habitats such as urban rivers, streams, wetlands, marshes, ponds, canals, flood plains, and coastal zones. These areas usually enjoy the highest regional biodiversity, and offer ecological services such as strong regulation of microclimate and water yield, recreation, and aesthetic value (Fig. 2.2).

(2) Semi-natural habitats in the urban parks and public green spaces

As the main urban biotic habitats, they include public green spaces such as parks and roadside green spaces, protected green buffers, as well as country parks, wildlife parks, botanical gardens, etc. (Fig. 2.3).

(3) Semi-artificial habitats in the attached green spaces

This type of habitats are the most accessible to urban residents, including recreational green spaces, community gardens, private courtyards, urban farms, ornamental lawns, and outdoor sport fields attached to lands used for living, business, office, culture, health care, sports, education, scientific research and design (Fig. 2.4).

(4) Artificial habitats on the sealed surfaces

The particularity of cities lies in the fact that they are home to numerous buildings, roads, and other sealed surfaces, which crowd out natural habitats. However, all these

Fig. 2.3 Semi-natural habitats in the urban parks and public green spaces (Left: Central Green park in Lujiazui, Shanghai; Right: Street Green Space in Shanghai Century Avenue)

Fig. 2.4 Semi-artificial habitats in the attached green spaces (Left: Green Space in Shanghai IFC; Right: Green Space in the Residential Area of Shanghai Huali Garden)

artificial buildings and structures contain heterogeneous microhabitats, which also provide new living surfaces and substrates for organic organisms within, above, and between their structures. Bridges, towers, buildings, electric wires, rails, and even building walls or roof surfaces that are dry and can withstand extreme day and night temperature variations accommodate cultivated or natural vegetation and ecological niches of invertebrates associated with them. On the layers of lime mortar in stone and brick walls grow xerophytes, native lithophytes, annuals, ferns and other plants that can tolerate the high calcium in these substrates. Many invertebrates such as insects, spiders, protozoa, and arthropods live on these plants. In the vertical direction, birds are easily adapt to the rockery-and cliff-like feature of high-rise buildings. Many birds and small animals build nests on human buildings. The artificial greening on roofs, potted plants on balconies, and vertical greenery also provide three-dimensional off-ground habitats for urban organisms. Although artificial habitats on the sealed surfaces cannot replace the lost the primitive nature habitats, they can help supplement and create new habitats for birds, bats, invertebrates, and multiple vegetation. The quality and type of an artificial surface also play an important role in defining its biodiversity. No animal or plant can survive on hard surfaces like concrete and

Fig. 2.5 Artificial habitats on the sealed surfaces (Left: Plants on the Ancient City Walls of Nanjing; Right: 3D Greenery of Shanghai Hongqiao Central Business District)

asphalt, which can also cause waterlogging through significant rainwater runoff, but surfaces made of pervious grit, floors paved with wooden grilles, and honeycomb walls planted with grass can create living spaces with great potential for small plants and invertebrates (Fig. 2.5).

(5) **Habitats with natural-succession on the abandoned land**

With the speedup of social changes and transformation, especially the rapid development of industrialization, urbanization, and modernization, the renewal in urban functions has led to a sharp increase in abandoned land (brownfields), including industrial brownfields, traffic brownfields, and mining brownfields. In the absence of human regulation and interference, vegetation on the brownfields is a great attraction to wildlife after spontaneous growth and natural succession. The abandoned land often has more species than surrounding green spaces, and even provides living spaces for rare plants and animals. For example, some abandoned lands have become hotspots for wild omnivorous animals after being reduced to landfills; and some low-lying lands near rivers have evolved into natural wetlands, becoming home to wild birds. Open spaces and brownfields provide unconventional physical changes to the soil, which help reinforce specific vegetation communities. In some countries, cemeteries are also often hotspots for the natural succession of many species (Fig. 2.6).

2.1.3 Patch Functions and Supply Potential of Each Type of Habitats

According to the principles of landscape ecology, patches are the most basic and suitable habitat for a certain biological population in the ecosystem. From the perspective of conservation biology, patches can be ecologically explained as "sources," "sinks," and "stepping stones." They are "sources" because the species inhabiting and breeding in a large resource patch (population) have such a large density and quantity that the organisms on it keep flowing outward. Known as the "sources" of

Fig. 2.6 Habitats with natural-succession on the abandoned land (Left: An Abandoned Factory in Dessau, Germany; Right: A small wetland at Shanghai Science and Technology Museum)

composite populations, they can continuously provide species to the surrounding smaller local populations. They are "sinks "as the local large population resource patches distributed around a large resource patch, with suitable habitats, will become convergence points for the organisms in the landscape, yet the species in those patches mainly live on species supplement and reimplantation from the large resource patch (Chen et al. 2016). "Stepping stones" are smaller transitional patches that help individual wildlife spread to distant places and facilitate the move of individuals between patches of various habitats (Fu et al. 2011).

According to the laws of nature, the basic needs of all wildlife include food, water, and shelter or cover (Johnston and Newton 1993). Adequate food is a must for wildlife to obtain sufficient nutrition for growth and to maintain a healthy body; water helps them digest, reduce their body temperature, and remove metabolic wastes; shelter protects them from bad weather and predators and improves their ability to survive or reproduce. Animals seek shelter for nesting, resting, escaping, raising their young, roaming, and withstanding heat and cold (Yarrow 2009).

Among the five types of habitats, near-natural habitats on the non-development land and semi-natural habitats in the urban parks and public green spaces are the largest in size and least disturbed by humans, and can thus become the "sources" to provide most urban wild animals with the space for nesting, feeding, and resting. Based on the size and quality of patches, Semi-artificial habitats in the attached green spaces can become either "sinks" or transitional "stepping stones" of different dimensions, mainly providing resting space and part of feeding space (it will be difficult for them to provide nesting space). Artificial habitats on the sealed surfaces have the potential to become "stepping stones," providing resting space and occasionally feeding and nesting space. Habitats with natural-succession on the abandoned land can provide abundant feeding space and some resting space, and whether they can provide nesting space depends on their qualities after natural-succession. They have the potential to become special "convergence points" or even "sources" in a city. The planning of the five habitats also corresponds to different planning control levels, as shown in Table 2.1.

Table 2.1 Types of Urban multiple biotic habitats, their functions and supply potential

Type	Degree of human-nature overlapping	Related land use	Size	Main functions	Main niches	Planning category
Near-natural habitats on the non-development land	Lowest	Farmland (N), waters & other unused land (E)	Large	Primary "sources"	Nesting, feeding, resting	Master plan
Semi-natural habitats in the urban parks and public green spaces	Not that Low	Green space (G)	Not that large	Secondary "sources"	Nesting, feeding, resting	Master plan, regulatory plan
Semi-artificial habitats in the attached green spaces	Not that High	Residential land (R), public facilities (C), etc.	Not that small	"sinks"/"stepping stones"	Resting, feeding, nesting	Regulatory plan, site plan, urban design
Artificial habitats on the sealed surfaces	High	Roads & squares (S), etc.	Small	"Stepping stones"	Resting	Site plan, urban design
Habitats with natural-succession on the abandoned land	Low	Industrial land (M), land for urban development (X), land for environmental sanitation facilities (U3)	Large & small	Potential "sources"/"sinks"	Feeding, resting	Regulatory plan

Note As the empirical studies in Chaps. 3 and 4 of this book take the case of Shanghai, the land use categories in the above table are subject to the classification of urban and rural land in the *Technical Guidelines for Controlled Detailed Planning of Shanghai* (2016 Revision)

2.2 Analysis Framework of the Relationship Between Urban Biodiversity and Built Environment

2.2.1 Characteristics of the Impact of Urban Built Environment on Biodiversity

The term "built environment" usually refers to the physical space and environment for human activities. It includes scales of buildings, parks, green spaces, streets, and even cities, as well as infrastructure such as water supply and energy network. It is a physical, spatial, and cultural product of human activities, and its size, spatial form, and other physical characteristics can be perceived by humans and other living things. The urban built environment only includes the living environment for humans in a narrow sense, but should be expanded to include the habitats for urban organisms if they are also regarded as users of the urban space environment. From the perspective of urban planning, this book pays more attention to such elements as land use, space configuration, and environmental development, which have a large impact on biodiversity and are easy to control in the process of urban planning. Examining the relationship between these elements and biodiversity will facilitate the adoption of corresponding strategies in urban planning. This book deals with the built environment at two scales: the macro scale which refers to larger cities, districts, or counties; and the meso-micro scale which refers to smaller built environments featuring homogeneous internal space and social attributes such as population, society, and economy. This book focuses on the residential areas, business areas, and office areas in the high-density central city, taking blocks as the statistical unit. It also covers specific micro-habitats and street green spaces at the micro scale.

The urban built environment is mainly composed of buildings, roads, pollutants, noises, vehicle flow, and pedestrian flow. Disorderly urban sprawl has caused the shrinkage of many wildlife habitats, directly threatening the reproduction of populations; and industrial pollutions and willful discharge of household garbage have altered the physiological characteristics and habitats of organisms, leading to the extinction or population decline of many species.

The impact of urban built environment on urban biodiversity is mainly reflected in the unique physiological and community adaptation characteristics of urban plants and animals. The physiological characteristics of urban plants include: low coverage; slow succession; long florescence; and strong anti-pollution ability. Community characteristics include: high proportions of cosmopolitan species, common species, and naturalized plants; herbaceous plants outnumbering woody plants; large proportions of weeds and companion plants; and flowering angiosperms usually being the dominant species (Forman 2013; Douglas and James 2014).

Compared with plants subject to more artificial interference, urban animals (wild animals, in particular) generally show synurbization during their evolution, i.e., non-domestic animals in cities gradually adapt to the artificial environment. Sometimes, their survival density is even higher in cities than that in the primitive nature, and they are more adaptable to urban environment (Luniak 2004; Francis and Chardwick 2012). This is especially true for birds and small rodents (Łopucki et al. 2013; Tomiałojć 2017). Compared with their counterparts in nature, urban wildlife show the following physiological characteristics: they are smaller in size, which makes it easy for them to move frequently; there is a high proportion of omnivorous animals, whose diet can change at any time, and some animals tend to eat what they are offered by humans, such as waste residues; they can build nests and roosts in artificial structures, with longer breeding periods, earlier maturation periods, and higher reproductive and survival rates; they can adapt to the high-density environment and have the behavior pattern of adapting to high-rise buildings, which are similar to the rock cliffs in the primitive nature; they are used to and even enjoy the interference of human activities, or adapt to humans behaviorally, and can physiologically tolerate great changes in non-biological conditions. For instance, the flush distance of birds is shorter; their activity time during day and night is longer, their range of activity is smaller, and they tend to migrate less. Urban wildlife show the following community characteristics: generalist species are in the dominant position, with one or several species being the main species in a city; specialist species play a minor role, making use of readily available food, shelter, and water; and there is a high proportion of cosmopolitan species (Forman 2013).

Synurbization also affects urban wildlife's selection of spaces. Take urban birds' nesting behavior for example, Wang et al. (2003) and Mikula et al. (2014) analyze and compare the density and height of bird nests in cities, their distances to city centers, to walkways and buildings, and to waters and woodlands, the canopy coverage, and the species of trees, and find that different species of birds have different distribution of nests, which indicates that birds adopt an adaptive strategy for selecting urban spaces.

2.2.2 *Factors in Urban Built Environment Affecting Biodiversity*

All living things on the Earth depend on the four environmental factors to survive: the sun, air, water, and soil. Those four factors underlie the Earth's ecosystem. Less affected by human activities, the sun and air are mainly biota factors; while water and soil are greatly threatened by human activities, and urban development is especially destructive to the water and soil environment. Most scholars believe that climate is the main factor affecting the biotic communities of a region; land use and landscape pattern play a decisive role in the biodiversity distribution pattern of a city; and on the meso-micro scale, micro-environmental factors such as habitat characteristics, vegetation, land development and intensity, and soil may be the key factors that determine the distribution of species (Mao et al. 2013). Therefore, the discussion on how urban built environment affects biodiversity in multiple habitats focuses on how it alters the natural waters and land resources, as well as their structures and environmental quality. Based on the idea of multiple overlapping habitats, the author explores the factors in the urban built environment affecting biodiversity, and concludes on the two dimensions: the habitat quality in the built environment and the pressure of human-induced disturbances on the natural environment. The former is about the substrate quality directly carrying the organisms, mainly including the factors in the urban built environment altering land resources and their structures and environmental quality; the latter is about the factors in the artificial environment indirectly disturbing natural species, patterns, and biological activities (Fig. 2.7).

(1) **Habitat quality: direct positive implications**

The habitat quality mainly includes the ecological land carrying plants and the plant carrying animals.

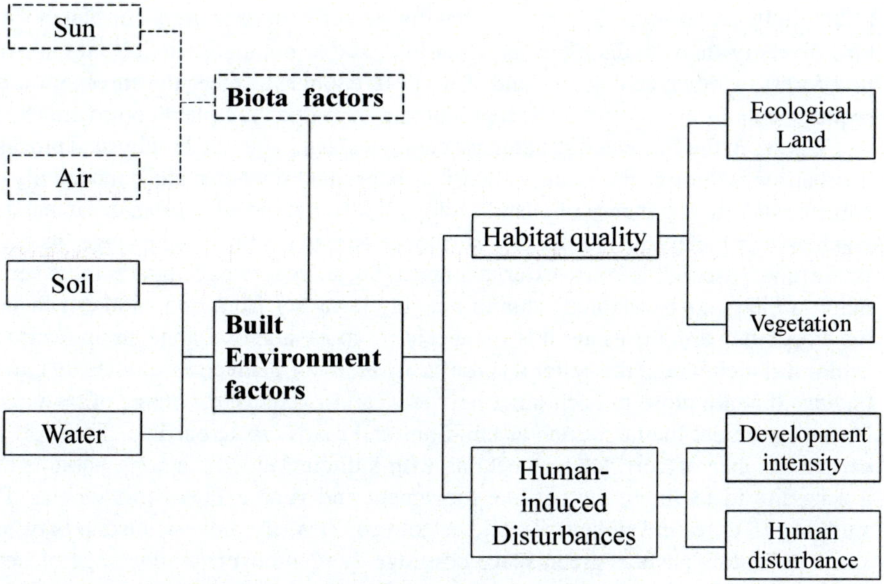

Fig. 2.7 Urban environment factors affecting Urban biodiversity

a. **Ecological land (habitat quality of plants)**

Ecological land refers to the land that mainly provides ecological services such as environmental regulation and biological conservation and plays an important role in maintaining regional ecological balance and sustainable development. It typically includes green space, woodland, wetland, cultivated land and waters. Changes in the function and structure of urban land, which can lead to shrinkage of ecological land, alterations in ecological land's patterns and forms and fragmentation of natural landscape, are the main reason for the changes in species structure and richness (Yan et al. 2012).

According to the "species-area relationship" theory in conservation ecology, there is a certain functional relationship between the number of species and the area in a certain region: generally speaking, the larger a habitat is, the more species it accommodates, and the healthier its flora and fauna become. Changes in urban land use directly lead to significant reduction of natural ecological space, especially that of forests and forest coverage, rendering habitats fragmented and unable to provide the minimum living area for wild animal populations. Take birds as an example. A study in Japan shows that dense forest birds can only be found in a well-vegetated green space with a minimum area of 1 hm^2, and large numbers of forest birds can only be found in a well-vegetated green space with a minimum area of 10 hm^2 (Lin 2001). Kathryn (1995) believes that the forest coverage and the width of forest land in an area exert profound impact on habitat fragmentation, birds' ability to survive and breed, changes in vegetation, and alien invasive plants. A study by Alberti et al.

(2007) in Puget Sound, USA, reveals that the area of a forest is more correlated with bird diversity than its distribution. Therefore, stable populations can only appear in a forest or other ecological land if it is big enough to meet the threshold area requirements for the survival and reproduction of animal and plant communities.

Changes in the space and structure of ecological land alter the biophysical process in urban areas through the location, density, shape, heterogeneity, and connectivity of habitat patches. The theory of island ecology holds that the edge of an urban habitat patch usually means the interference surface of human activities. The longer its edge is, the more susceptible it is to external impact. So, an integrated habitat is more beneficial to biodiversity. Habitats should not stay far away from one another because the closer they are, the easier it is for species to move and exchange among various ecological niches, and the better it is for the diversity of plant communities. In cities, habitats that are close to each other may still fail to achieve exchange of resources because artificial facilities such as buildings and roads create barriers. Therefore, a certain number of corridor connections with sufficient widths among habitats are a necessity to facilitate the net-like movement and gene exchange of species. By studying 15 cities in England, Tratalos et al. (2007) verify the relationship between space indicators such as green space coverage, garden coverage, coverage of trees in gardens and green spaces, average area of green space patches, average area of gardens and green space patches, average area of non-enclosed land patches, and average area of tree patches in gardens and green spaces, and biodiversity potential and the service indicator of ecosystem. Reis et al. (2012) examine the correlation between 20 variables related to vegetation, land coverage, and land use types, and the bird species richness in eight blocks, and concludes that the proportion of residential land, the density of native trees, and the proportion of unpaved roads are the variables with the most positive correlation, while the density of commercial land, the density of exotic trees, and the ratio of built area show the most negative correlation. Through their study of 29 urban parks in Beijing, Xie et al. (2016) examine how the landscape characteristics of a park (including the park area, landscape fragmentation, proportions of woodland and artificial surface, maximum woodland patch index, landscape aggregation index, landscape diversity, and evenness index) influences the avian community during the summer months, and find that as the average patch area is small in a park, landscape diversity and evenness index have a significant adverse effect on the species richness of birds.

b. **Vegetation (carrying elements of animal)**

Urban vegetation directly provides food and shelter for wild animals, which feed on the stamens, fruits and seeds of many fruit trees and shrubs all the year round. Trees with high canopy density can protect wildlife from human disturbance. Tall, long-lived trees are often home to many species, and their lumens make good nests. Water from fruits and berries, as well as morning dews and raindrops on plants and lawns, are also indirect water sources for wildlife. Changes in urban vegetation and its coverage area, proportion, and biomass, as well as different species structure and distribution, will affect the spatial niches where wild animals feed, nest, and inhabit, thereby affecting the number and distribution of species.

The decrease of vegetation coverage is the main reason for the decline in urban biodiversity. For animals in a certain area, the species richness is significantly positively correlated with the vegetation coverage. Lee et al. (2004) study the birds in major cities in Taiwan and find that bird richness and normalized difference vegetation index (NDVI) are significantly positively correlated. Parker and Nilon (2012) study the grey squirrels at six parks in Baltimore and prove that park area, vegetation coverage, and the number of tree arrays are the three major indicators affecting the density of grey squirrels. Arbor, shrub, and grass vegetation layers provide different food and shelter substrates. Ge et al. (2005) study the birds in eight garden green spaces in Shanghai during the spring months, and conclude that factors such as the number of vegetation species, the coverage of shrub layers, and the coverage of herb layers play a critical role in the structure and distribution of the avian community in a garden.

In general, the more abundant the vegetation, the more abundant the wildlife diversity; and the less diverse the vegetation, the less diverse the species. Through his study of the relationship between plant structure and insect diversity in the green spaces of Mianyang, Sichuan, Xiao (2005) confirms that the distribution of insect community is not only influenced by the number of plants, but also by the layered structure, temperature, and humidity of the green spaces. Yan et al. (2006) show in their study of butterflies in Chongqing that butterflies are more abundant in species, larger in population, and greater in diversity on those habitats with richer vegetation species and larger coverage. The decrease of native vegetation in cities will inevitably lead to a sharp deterioration in the habitats for wild animals. Burghardt et al. (2009) conduct a study on six suburban properties in southeastern Pennsylvania, and find that compared with the control group, the properties where native vegetation takes predominance boast more lepidopteran species, higher avian species richness, higher biomass and more native breeding bird species.

The species diversity of wildlife is also related to the spatial arrangement of vegetation. Wang et al. (2003) study the breeding birds and their nests in the street trees of Hangzhou, and find that canopy coverage, distances to large woodlands and to main water sources, and tree belt width are positively correlated with the nest distribution of certain bird species. Studies by Van Heezik et al. (2008), and Ortega-Alvareza and MacGregor-Fors (2009) respectively indicate that the avian species richness in a New Zealand city and the southwest of Mexico City are positively correlated with the average height of trees, the coverage and height of shrubs, and the height of herbaceous plants.

(2) **Human-induced disturbances (indirect negative implications)**

The disturbances of biodiversity by the artificial environment are mainly manifested in the competition between urban land development activities and biotic habitats for various ecological niches, and the changes in the physiological and community adaptation characteristics of organisms brought by highly intensive human activities in cities.

a. **Development intensity (disturbances of habitats)**

Development intensity represents the degree of land use per unit, and usually includes FAR, building density, road network density, residential density, etc. Habitat isolation caused by human development activities has changed the natural process of biological survival and reproduction (e.g., pollen spread is blocked and animals are prevented from crossing). This can lead to the suppression of gene flows, increasing the possibilities of inbreeding and rare genetic variant loss, and affecting the process and trend of biodiversity changes, including the reproduction, growth, and development of species, as well as inter-species relationship (Wu and Lv 2008). Glista et al. (2009) argue that the occupation of biotic habitats by a traffic corridor led to the fragmentation and islandization of the habitats, preventing organisms on both of its sides from crossing and even causing unnatural deaths of wild animals. Ortega-Alvareza and MacGregor-Fors (2009) study the relationship between green space, residential land, commercial-residential land, and commercial land, and avian diversity and its spatial distribution in the urban areas of Mexico City, and find that the degree of urban development is positively correlated with the lack of diversity in the avian population structure. A study by Wang et al. (2012) on all districts and counties of Shanghai shows that human activity indicators, such as per capita GDP, population density, and traffic network density, have a negative impact on the overall biodiversity index and species richness.

b. **Human activities (disturbances of organisms)**

Vehicle flow, pedestrian flow, the noise, light, water, air, and soil pollution, and microclimate changes affect the biological rhythm and florescence of plants and the behavior patterns of animals in traveling, foraging, nesting, and communicating, thus leading to changes in physiological functions and population structure. Hu et al. (2005) point out that road traffic is the chief killer of wild vertebrates. Alberti and Marzluff (2004) and Miltner et al. (2004) find that the increase in urban impermeable surfaces alters surface hydrology and microclimate, leading to a decline in the biodiversity of subaqueous organisms and the total number of stream macroinvertebrates. Through their study of the relationship between the lighthouses at Lake Erie, Ontario, Canada and the deaths of nocturnal migratory birds, Jason and Charles (2003) find that changing the intensity of light beam could significantly reduce the misleading effect of light on birds and thus reduce their mortality. Urban human activities, such as conversion into conduits, cutting-off, cement lining, and building of stone revetment and high dykes, damage and pollute waters, can also lead to sharp drop in the species of birds that live on waters (Li et al. 2005) (Table 2.2).

As shown in the above table, efforts to protect and enhance urban biodiversity through optimizing the built environment should be made on the two dimensions: improving habitat quality and mitigating human-induced disturbances, which involve the four factors in the built environment: ecological land, vegetation, development intensity, and human activities. The factors in the urban built environment affecting biodiversity on the macroscale are the intensity, density, and agglomeration of overall

Table 2.2 Impact of Urban environment factor changes on organisms

Urban environment factors & characteristics	Major effects on plants	Major effects on animals
Climate: High temperature, low humidity, low wind speed, weak circulation	Longer growth cycle and changed phenological phase	Early breeding, accelerated breeding of some species, and disturbance of biological rhythm
Soil: High compactness, poor permeability, high sealing ratio, weak fertility, serious pollution	Lack of diversity in community structure and high proportion of invasive species	Changed animal community structure in the soil
Light: Less natural light, more artificial light, night light interference, daytime light pollution	Affecting the biological rhythm and florescence of plants	Changes in migration, hibernation, breeding, and foraging of nocturnal animals, and deaths caused by near-ground crash
Sound: Noise pollution	Early withering	Affecting the foraging ability of birds, bats and other groups that mainly rely on sound for communication
Water: Low underground water level, water pollution	Proliferation of water plants that are resistant to pollution	Affecting the species and populations of fish and benthic fauna
Air: Air pollution (dust/inhalable particulate matter, sulfur dioxide, nitrogen oxide, carbon monoxide, etc.)	Reduction or extinction of sensitive species, and preservation of species highly resistant to pollution	Slowing down the normal development of organisms, and affecting their physiological functions

development, and thc size and structure of various types of ecological land such as green space, woodland, wetland, cultivated land, and waters, but it is very hard to collect the data of vegetation on the macroscalc. On the meso-micro scale, apart from the land development intensity that is also one of the factors on the macroscale, green space and waters, which are regarded as small patches of ecological land in the central city, also have a direct influence on urban biodiversity. Therefore, we suggest that an empirical study can explore the relationship between the size and space characteristics of green spaces and waters and their biodiversity by focusing on green spaces and touching on waters. Additionally, plants can provide foraging, nesting and resting ecological niches for animals, and their size, structure and planting form can be surveyed, mapped and also analyzed as the built environment factors affecting biodiversity. In the field of urban planning, the impact of environment pollution is generally not considered as a key variable in studies of factors affecting biodiversity as it is hard to obtain environment monitoring data and directly regulate environment pollution in urban spatial planning.

2.3 The Hypnosis of the Influence Mechanisms of the Built Environment on Urban Biodiversity

Drawing on the theory of island biogeography put forward by Robert H. MacArthur and Edward O. Wilson in 1967, which analyzes the shape, area, isolation, and the permutation and combination of island habitats (Robert et al. 1967), this book proposes, based on the resource limitation and vertical extensibility of high-density urban spaces, the following six effects in the built environment on urban biodiversity in multiple habitats.

(1) **Ratio Effect**: As "aborigines" of cities, wildlife have been crowded out by reinforced concrete buildings and lost their original habitats. The process of urban sprawl is a process of increasingly fragmented habitats for wildlife. In high-density cities with limited land resources, the ratio resulting from the overlapping of biotic habitats and human settlements on different dimensions is the primary factor to determine the size of a biological substrate.

(2) **Area Effect**: According to the theory of "species-area relationship" in conservation ecology, there is a certain functional relationship between the number of species and the area in a certain region. Generally speaking, the larger a habitat is, the more species it accommodates, and the healthier its flora and fauna become.

(3) **Edge Effect**: The edge of an urban habitat usually means the interference surface of human activities. The longer its edge is, the more susceptible it is to external impact. Therefore, habitats complete in shape are beneficial to biodiversity. In other words, a circular habitat is better than a square one with the same area, and a square habitat is better than a rectangular one with the same area.

(4) **Distance Effect**: According to the theory of spatial niche approach, the closer several habitats are, the easier it is for species to migrate and communicate among various ecological niches, and the better it is for the diversity of plant communities. So, habitats should not be far apart from one another.

(5) **Network Effect**: In cities, habitats that are close to each other may still fail to achieve exchange of resources because artificial facilities such as buildings and roads create barriers. Therefore, a certain number of corridor connections with sufficient widths among habitats are a necessity to facilitate the net-like movement and gene exchange of species. In cities, green corridors, rivers, and green spaces often serve as the important green network.

(6) **Height Effect**: In high-density cities, there is a three-dimensional dynamic balance between natural habitats for organisms and the human settlements. In addition to the Earth's surface, it also includes water, soil, and vertical space. Various species and humans take up the three-dimensional spatial niches through dislocation and overlapping of time and space. The height effect can not only preserve more ground green space for biotic habitats, but also provide more sealed surfaces as artificial habitats.

2.3.1 Impact Mechanism of Urban Built Environment on Biodiversity in Urban Multiple Habitats

(1) Near-natural habitats on the non-development land

Near-natural habitats on the non-development land are affected first and foremost by the "human-organism" spatial ratio brought about by urban development. Large areas of habitat land and vegetation can provide near-natural habitats with a more stable community structure and a higher biodiversity. Large in area, near-natural habitats are far away from each other in the highly fragmented urban landscape, so distance effect can hardly play a role in them. Yet a corridor network formed by urban ecotones such as riparian zone, shore belt, and green road can strengthen the role of near-natural habitats as "sources," helping species spread outwards along the corridor. Height effect has no influence on near-natural habitats either because of the small number of constructions in them.

(2) Semi-natural habitats in the urban parks and public green spaces

Semi-natural habitats in the urban parks and public green spaces are also affected by the ratio effect and the size of habitat land and vegetation. The distance between park green spaces and the corridor network effect make it possible for biotic communities to move and exchange genes among large urban habitats. As park green spaces are usually complete, the edge effect will be excluded. The height of vegetation can bring more vertical ecological niches, which will benefit the diversity of species such as birds that are good at utilizing off-ground space.

(3) Semi-artificial habitats in the attached green spaces

As semi-artificial habitats in the attached green spaces are located in the major functional areas of a city, they are affected by both the ratio effect and the height effect resulting from urban development. The ratio effect determines the proportion of habitat space that can be used by organisms; and the height effect causes disturbance to the inhabitation of organisms while increasing the available space above the ground. Semi-artificial habitats in the attached green spaces greatly overlap with human settlements, so the interference effect of edge surface should be a key consideration. Meanwhile, the distance between and the network of habitat patches, as well as the size and height of vegetation also exert significant influence on the biodiversity of semi-artificial habitats.

(4) Artificial habitats on the sealed surfaces

Artificial habitats on the sealed surfaces almost completely overlap with human habitats, and the height effect also brings them both positive and negative effects. Their biodiversity mainly depends on the size of habitats on the sealed surfaces and their vegetation substrates.

Table 2.3 Impact mechanism of built environment in Urban multiple habitats

Type of habitats	Ecological land	Vegetation	Development intensity	Overall impact
Near-natural habitats on the non-development land	A + N	A	R	D + N
Semi-natural habitats in the urban parks and public green spaces	A + D + N	A + H	R	
Semi-artificial habitats in the attached green spaces	A + E + D + N	A + H	R + H	
Artificial habitats on the sealed surfaces	A	A	H	
Habitats with natural-succession on the abandoned land	A	A	–	

Note R—Ratio Effect, A—Area Effect, E—Edge Effect, D—Distance Effect, N—Network Effect, H—Height Effect

(5) Habitats with natural-succession on the abandoned land

Habitats with natural-succession on the abandoned land usually exist in cities as individuals, and their biodiversity is mainly determined by the size of land free from human disturbance and that of natural succession vegetation.

In addition to the impact mechanism of the built environment within the five types of habitats, the distance between multiple habitats and the network effect also play an important role. The systematic habitat networks as "sources," "sinks," "corridors," and "stepping stones" can improve both the biodiversity within various habitats and the biodiversity in general (Table 2.3).

References

Alberti M, Booth D, Hill K et al (2007) The impact of urban patterns on aquatic ecosystems: an empirical analysis in puget lowland sub-basins. Landsc Urb Plan 80(4):345–361

Alberti M, Marzluff JM (2004) Ecological resilience in urban ecosystems: linking urban patterns to human and ecological functions. Urb Ecosyst 7:241–265

Bateman IJ et al (2011) National ecosystem assessment- understanding nature's value to society: Technical Report. United Nations Environment Program World Conservation Monitoring Centre (UNEP-WCMC)

Breuste J, Pauleit S, Haase D et al (2016) Stadtökosysteme Funktion. Springer-Verlag GmbH, Berlin & Heidelberg, Management und Entwicklung

Burghardt TK, Tallamy WD, Shriver WG (2009) Impact of native plants on bird and butterfly biodiversity in suburban landscapes. Conserv Biol 23(1):219–224

Chen LD et al (2016) Analysis and application of source-sink landscape pattern. China Science Publishing & Media Ltd., Beijing (in Chinese)

Douglas I, James P (2014) Urban ecology: an introduction. Routledge, Abingdon & New York, pp 222–228, 241–242

Forman RTT (2013) Urban ecology: science of cities. Cambridge University Press, New York, pp 220–223, 241

Francis RA, Chardwick AM (2012) What makes a species synurbic? Appl Geogr 32:514–521

Fu BJ, Chen LD, Ma KM (2011) Principles and applications of landscape ecology(2nd Version). China Science Publishing & Media Ltd., Beijing (in Chinese)

Ge ZM, Wang TH, Shi WY et al (2005) Impacts of environmental factors on the structure characteristics of avian community in shanghai woodlots in spring. Zool Res 26(1):17–24 (in Chinese)

Gilbert OL (1991) The ecology of urban habitats. Chapman & Hall, London

Glista DJ, DeVault TL, DeWoody JA et al (2009) A review of mitigation measures for reducing wildlife mortality on roadway. Landsc Urb Plan 91(1):1–7

Han XL, Li DH (2009) Progress in the study on urban remnant seminatural habitats. J Nat Resour 24(4):561–566 (in Chinese)

Hu ZJ, Yu CQ, Xu HF et al (2005) Ecological effects of roads on terrestrial animals. Chinese J Ecol 24(4):433–437 (in Chinese)

Jason J, Charles MF (2003) The effects of light characteristics on avian mortality at lighthouses. J Avian Biol 34:328–333

Johnston J, Newton J (1993) Building green. London ecology unit

Kathryn F (1995) Assessing effects of agriculture on terrestrial wildlife: developing a hierarchical approach for the US EPA. Landsc Urb Plan 31(1–3):99–115

Köstler H,Grabowski C, Moeck M, et al (2003) Beschreibung der biotoptypen Berlins.Berlin,Germany: Arbeitsgemeinschaft Biotopkartierung, Senatsverwaltung für Stadtentwicklung und Umwelt. http://www.berlin.de/senuvk/natur_gruen/naturschutz/biotopschutz/download/biotopkartierung/biotoptypen.pdf. Accessed 1 Sept 2016

Kowarik I (1992). Das Besondere der städtischen Flora und Vegetation. In: Natur in der Stadt -der Beitrag der Landespflege zur Stadtentwicklung. Schriftenreihe des Deutschen Rates für Landespflege, 61. Aufl, S 33–47

Lee PF, Ding TS, Hsu FH et al (2004) Breeding bird species richness in Taiwan: distribution on gradients of elevation, primary productivity and urbanization. Biogeography 31:307–314

Li DH (2016) Fragmentation-the biggest obstacle for biodiversity conservation. Landsc Archit Front 3:34–39 (in Chinese)

Li JS, Gao GX, Zhang XL et al (2005) Effects of urbanization on biodiversity: a review. Chinese J Ecol 24(8):953–957 (in Chinese)

Lin XD (2001) Urban and rural ecology, 2nd edn. Chan's Arch-Publishing Co., Ltd, Taipei (in Chinese)

Łopucki R, Mroz I, Berliński L et al (2013) Effects of urbanization on small-mammal communities and the population structure of synurbic species: an example of a medium-sized city. Can J Zool 91(8):554–561

Luniak M (2004) Synurbization-Adaptation of animal wildlife to urban development. In: Shaw et al (eds) Proceedings of the 4th International Urban Wildlife Symposium, Tucson, AZ, USA, pp 50–55

Mao QZ, Ma KM, Wu JG et al (2013) An overview of advances in distributional pattern of urban biodiversity. Acta Ecol Sin 33(4):1051–1064 (in Chinese)

Mikula P, Hromada M, Albrecht T et al (2014) Nest site selection and breeding success in three turdus thrush species coexisting in an urban environment. Acta Ornithol 49(1):83–92

Miltner RJ, White D, Yoder C (2004) The biotic integrity of streams in urban and suburbanizing landscapes. Landsc Urb Plan 69(1):87–100

Müller N, Ignatieva M, Hilon CH, et al (2012) Patterns and trends in urban biodiversity and landscape design. In: UN-Habitat et al (eds) Urbanization, biodiversity and ecosystem services: challenges and opportunities-a global assessment. Nairobi: UN-Habitat, pp 123–174

Ortega-Alvareza R, MacGregor-Fors I (2009) Living in the big city: effects of urban land-use on bird community structure, diversity, and composition. Landsc Urb Plan 90(3–4):189–195

Parker ST, Nilon HC (2012) Urban landscape characteristic correlated with the synurbanzation of wildlife. Landsc Urb Plan 106:316–325

Pickett STA, Cadenasso ML, Grove JM et al (2001) Urban ecological systems: linking terrestrial ecological, physical, and socioeconomic components of metropolitan areas. Annu Rev Ecol Syst 32:127–157

Reis E, López-Iborra GM, Pinheiro RT (2012) Changes in bird species richness through different levels of urbanization: implications for biodiversity conservation and garden design in Central Brazil. Landsc Urb Plan 107:31–42

Robert H, MacArthur Edward O et al (1967) The theory of island biogeography. Princeton University Press, Princeton

Tertilt TM (2010) Enhancing Urban biodiversity: guiding principles for ecological landscaping. Center for Urban Greenery & Ecology, Singapore National Park Board

Tomiałojć L (2017) Human Initiation of Synurbic Populations of Waterfowl, Raptors, Pigeons and Cage Birds. In: Murgui E, Hedblom M (eds) Ecology and conservation of birds in Urban environments. Springer, Cham, pp 271–286

Tratalos J, Fuller RA, Warren PH et al (2007) Urban form, biodiversity potential and ecosystem services. Landsc Urb Plan 83:308–317

Van Heezik Y, Smyth A, Mathieu R (2008) Diversity of native and exotic birds across an urban gradient in a New Zealand city. Landsc Urb Plan 87:223–232

Wang Q, Ruan JJ, Sha CY et al (2012) Human impacts on the spatial pattern of biodiversity of Shanghai City. Ecol Environ Sci 21(2):279–285 (in Chinese)

Wang YP, Chen SH, Ding P (2003) Breeding birds and their nests in street tree strips in Hangzhou City. Zool Res 24(4):259–264 (in Chinese)

Whittaker RH, Levin SA, Root RB. Niche, habitat and ecotope (1973) American Naturalist 107(955): 321–338

Wittig R (1998) Flora und Vegetation. In: Sukopp H, Wittig R (eds) Stadtökologie, 2nd edn. Gustav Fischer Verlag, Stuttgart, pp 219–265

Wu JG, Lv JJ (2008) Effects of land use change on the biodiversity. Ecol Environ 17(3):1276–1281 (in Chinese)

Xiao K (2005) Study on the relationship between different types of green space and insect diversity in Urban gardens in Mianyang. Xiandai Hortic 10:10–11 (in Chinese)

Xie SL, Lu F, Cao L et al (2016) The effect of landscape patterns on Avian communities during summer months in Beijng's urban parks. Landsc Archit Front 3:10–21 (in Chinese)

Yan H, Yuan XZ, Liu WP et al (2006) Butterfly diversity along a gradient of urbanization: Chongqing as a case study. Biodivers Sci 14(3):216–222 (in Chinese)

Yan WT, Xiao JH, Hu H et al (2012) Urban spatial structure and environmental performance: review and thought. Urb Plan Forum 2:50–59 (in Chinese)

Yarrow G (2009) Habitat requirements of wildlife: food, water, cover and space. For Nat Resour Fact Sh 14:1–5

Chapter 3
Relationship Between Urban Biodiversity and Built Environment and in Macro Scale: The Case Study of Shanghai, China

3.1 Urbanisation and Urban Biodiversity in Shanghai

As the metropolitan city with the highest urbanisation rate in China, Shanghai is a coastal city located at the middle part of the North–South coastline in China and the estuary of Yangtze River with rich terrestrial–rivershed–marine ecotones that provide diverse types of habitat and food resources for wildlife. Apart from accommodating over 24 million inhabitants in 7 urban central districts and 9 suburban districts within its 6833 km^2 territory (Fig. 3.1), Shanghai is also home to more than 780 species of wild vascular plants, 300 species of freshwater fish and 530 species of terrestrial vertebrates, including 14, 36, 445 and 42 species of amphibians, reptiles, birds and mammals, respectively (Shanghai Environmental Protection Bureau et al. 2013).

However, as a consequence of its rapid urban development over the last four decades, the built-up area of Shanghai has rapidly increased along with its land use intensity and population density, thereby limiting its urban natural ecological space. From 2006 to 2008, the ecological land area (i.e. non-built-up land including green spaces, forest lands, cultivated areas, waters and wetlands, especially coastal wetlands) of Shanghai decreased from 4462 km^2 to 4057 km^2, which translates to a continuous reduction of habitat space for wildlife (Lu et al. 2016). Human activities are also evident in the remaining natural habitats, which lead to a phenomenon known as 'habitat fragmentation' where terrestrial and aquatic animal populations are kept at low levels as well as demonstrate isolated distribution and threatened genetic diversity (Shanghai Environmental Protection Bureau et al. 2013). The newly released Shanghai Master Plan (2017–2035) has focused on green development by exploring a new growth model that is suitable for promoting the ecological development of densely populated megacities. A green and open eco-network is proposed to serve as an ecological security barrier that protects biodiversity and improves the quality and stability of the ecosystem (Shanghai Urban Planning and Land Resource Administration Bureau 2018). Therefore, this plan is expected to promote a balance between human settlement and natural habitat and to protect biodiversity in a new round of urban development.

J. Gan, *Urban Biodiversity and Built Environment*,
https://doi.org/10.1007/978-981-16-0949-7_3

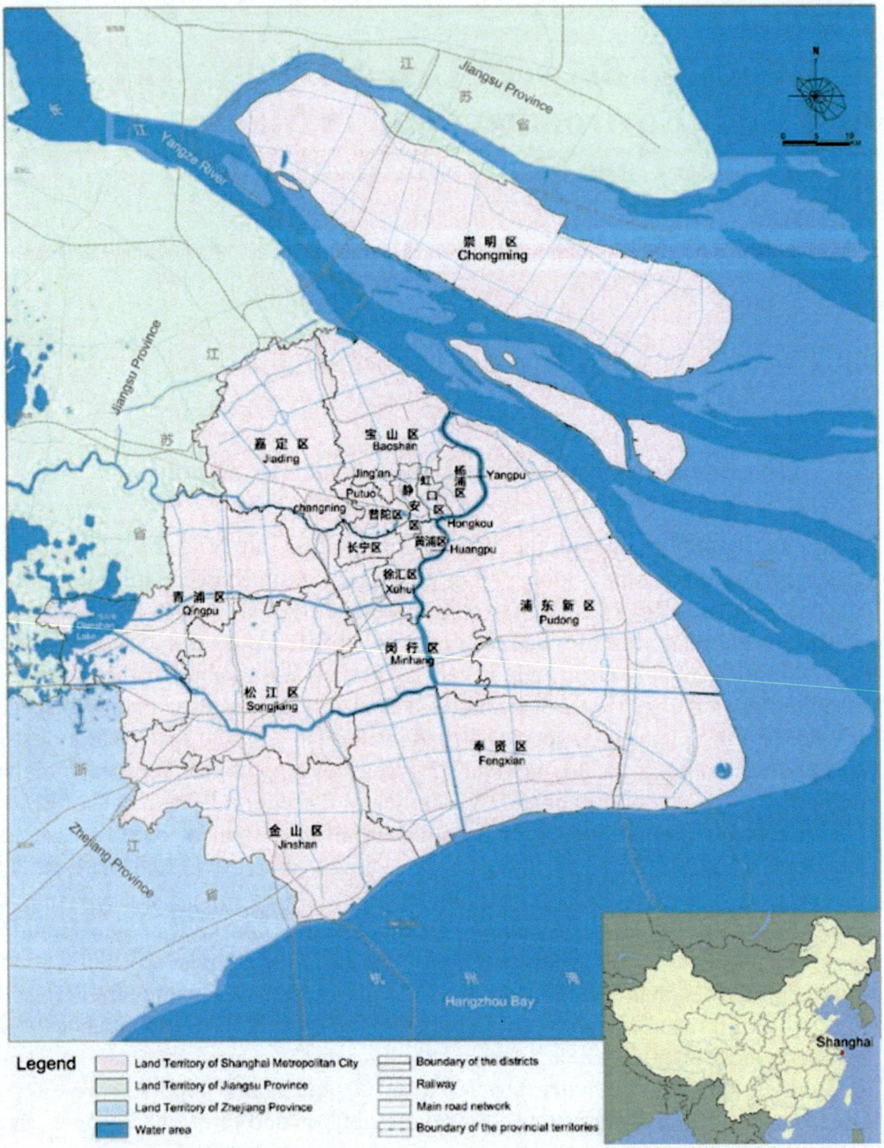

Fig. 3.1 Location of Shanghai in China and its distinct administrative units. *Note* The figure is translated and redrawn based on the *Map of Land Territory of Shanghai Metropolitan City* from the *Atlas of Shanghai Master Plan (2017–2035)* (Shanghai Urban Planning and Land Resource Administration Bureau 2018)

3.2 Materials and Methods

3.2.1 Data on Biodiversity and Its Spatial Distribution

Statistics on the biodiversity of each district of Shanghai were retrieved from the *Assessment Report for Biodiversity Based on Species Richness of Wildlife and Plants* sponsored by the Ministry of Environmental Protection and undertaken by the Shanghai Institute of Environmental Sciences in 2011. In this report, district-level administrative areas were used as research units, with urban central districts (e.g. Yangpu, Hongkou, Huangpu, Jingan, Changning, Xuhui and Putuo) acting as a single statistical unit and suburban districts (e.g. Baoshan, Qingpu, Minghang, Jiading, Songjiang, Jingshan, Fengxian, Pudong and Chongming) acting as distinctive statistical units (Wang et al. 2012). Based on the retrievable literature on recorded fauna and flora species, data on species richness were summarised and an integrated biodiversity index (*BI*)[1] was calculated according to the *Standard for the Assessment of Regional Biodiversity* (HJ623-2011) (Ministry of Environmental Protection of People's Republic of China 2011). The *BI* and species richness data for each spatial unit are shown in Figs. 3.2, 3.3, 3.4, 3.5, 3.6, 3.7, 3.8, 3.9 and 3.10.

3.2.2 Built Environment Variables and Calculation Method

Three types of built environment variables that influence biodiversity were selected for this study.

(1) **Development intensity**

Urban development has important effects on biodiversity. In metropolitan areas, economic development and the consequent population aggregation are important factors that threaten biodiversity. Four variables that evaluate the intensity of human development activities, namely, economic development, population density, built environment density and building intensity, were selected in this study, whereas GDP per capita, GDP per square meter, population density, road network density, building density and proportion of high-rise buildings were used as indicators. Most of the data were obtained from the *Shanghai Statistical Yearbook 2012* (Shanghai Statistics Bureau 2012). Given that data on the floor area ratio (FAR) of each district are unavailable, we used data on the area of high-rise buildings (buildings with over 8 floors) from the *Shanghai Statistical Yearbook 2012* as a proxy to calculate the area

[1]Note: Following the *Standard for the Assessment of Regional Biodiversity* (HJ623-2011), *BI* was calculated as $BI = R_v{}'*0.2 + R_p{}'*0.2 + D_E{}'*0.2 + E_D{}'*0.2 + R_T{}'*0.1 + (100\text{-}E_I{}')*0.1$, where $R_v{}'$ denotes the normalised richness of wild animal species, $R_p{}'$ denotes the normalised richness of wild vascular plant species, $D_E{}'$ denotes the normalised diversity of various types of ecosystems, $E_D{}'$ denotes normalised species endemism, $R_T{}'$ denotes the normalised richness of threatened species and $E_I{}'$ denotes the normalised degree of invasion of invasive alien species.

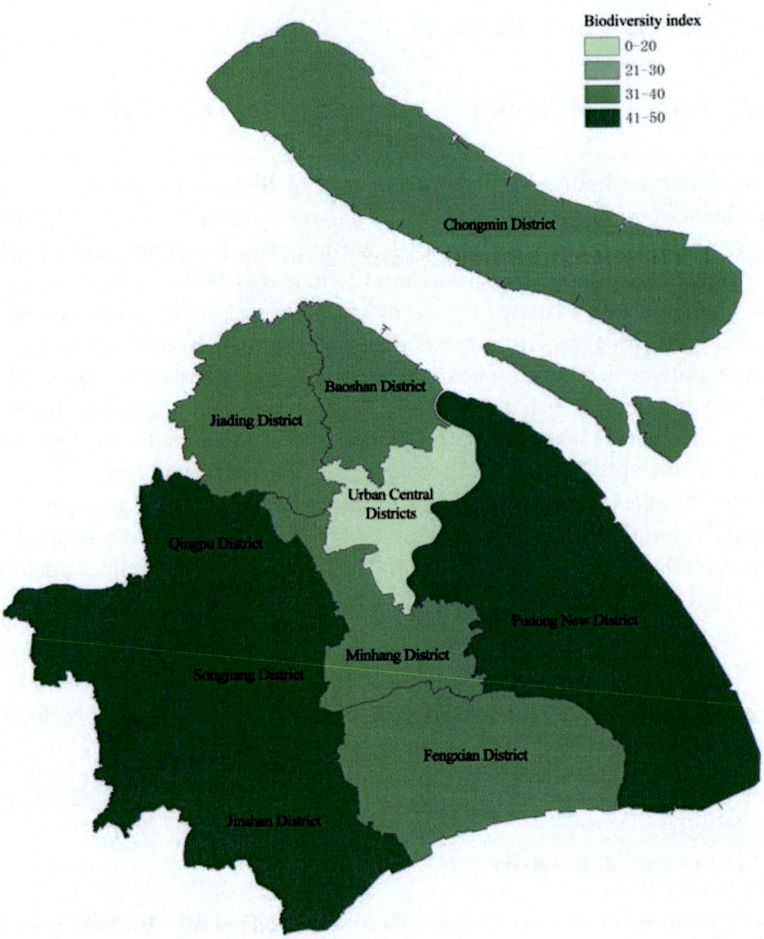

Fig. 3.2 Biodiversity index for each district of Shanghai (2011)

of high-rise buildings and the gross floor area of buildings with more than 8 floors (over 24 m high), more than 16 floors (over 50 m high) and more than 30 floors (over 90 m high). We used these variables as proxies to FAR and to characterise building intensity. The data for road network density were obtained by using the spatial analysis tool of ESRI ArcGIS10.2.

(2) **Ecological land**

a. **Area and proportion of ecological land**

Ecological land has received much attention in urban development projects given its role in maintaining ecological safety, protecting biodiversity, and providing other

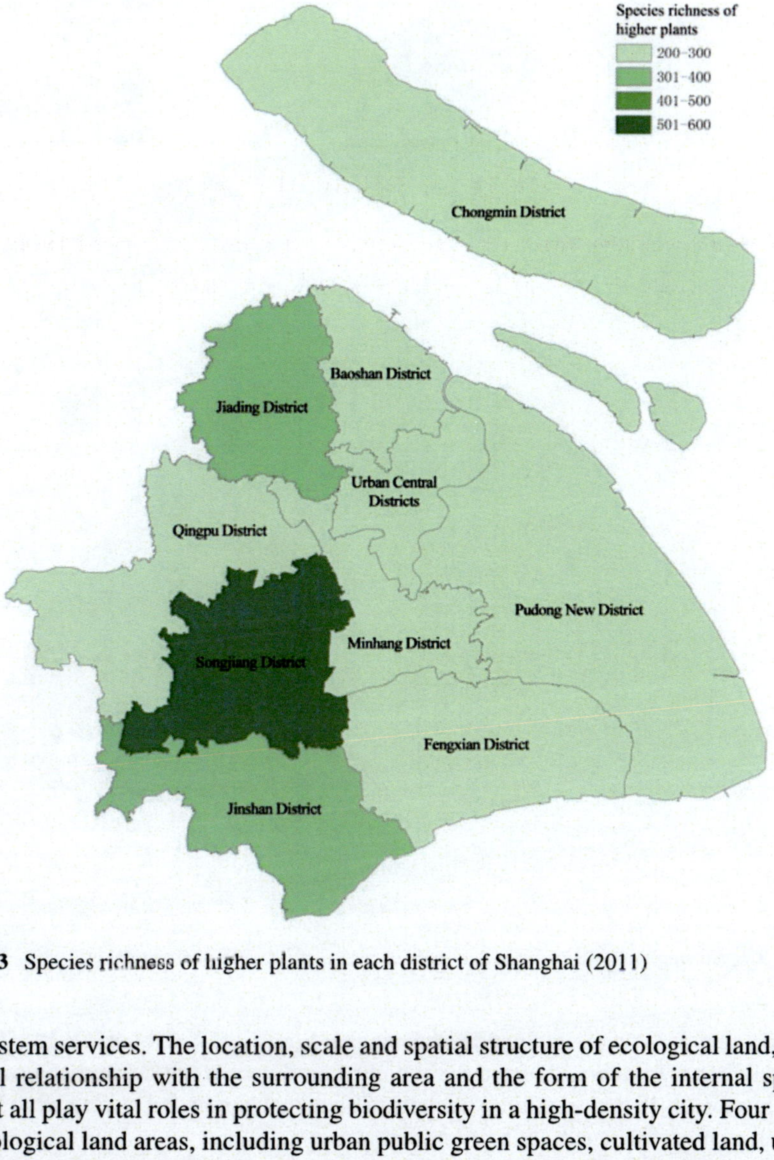

Fig. 3.3 Species richness of higher plants in each district of Shanghai (2011)

ecosystem services. The location, scale and spatial structure of ecological land, their spatial relationship with the surrounding area and the form of the internal spatial layout all play vital roles in protecting biodiversity in a high-density city. Four types of ecological land areas, including urban public green spaces, cultivated land, urban waters and wetlands and forest lands, were selected for the analyses. The total area and the proportion of the occupied area of these four types of ecological lands were used as variables.

Data on the ecological lands of each district of Shanghai were obtained by Yang (2016) from a satellite imagery of the city captured in 2013 and by calculating the proportion of each ecological land based on the internal administrative borders of the city. Data on the park area (a type of urban public green space) and average area

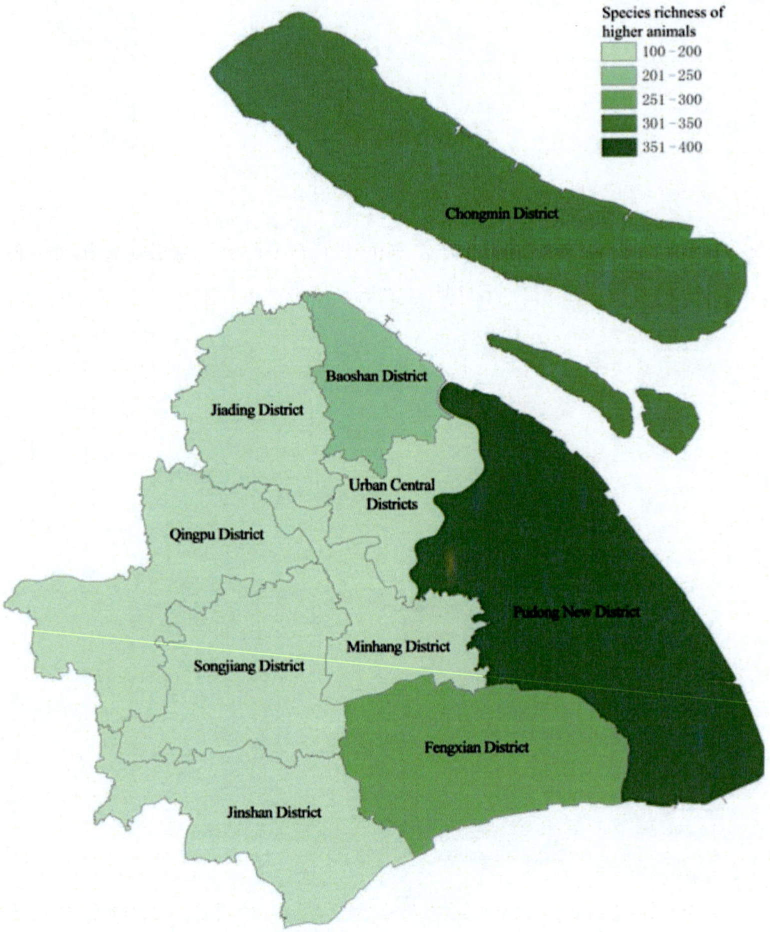

Fig. 3.4 Species richness of higher animals in each district of Shanghai (2011)

of a single park were obtained from Shanghai Statistical Yearbook (2012). The data of wildlife habitat area is from the report of biodiversity survey and environmental evaluation on 45 major wildlife habitats in Shanghai conducted by Shanghai Wildlife Conservation Management Station (2008).

b. **Water network**

The remaining water courses in urban areas become ecological corridors. The landscapes formed by riverbanks and adjacent non-built land are generally known as the most abundant ecotones for animal and plant species as well as the best corridors for fish, amphibians, reptiles, birds, and other wildlife that rely on these landscapes to migrate, mate and exchange genes. Therefore, we used water network as a variable of a macroscale spatial form. Following Cantwell and Forman (1993), we derived

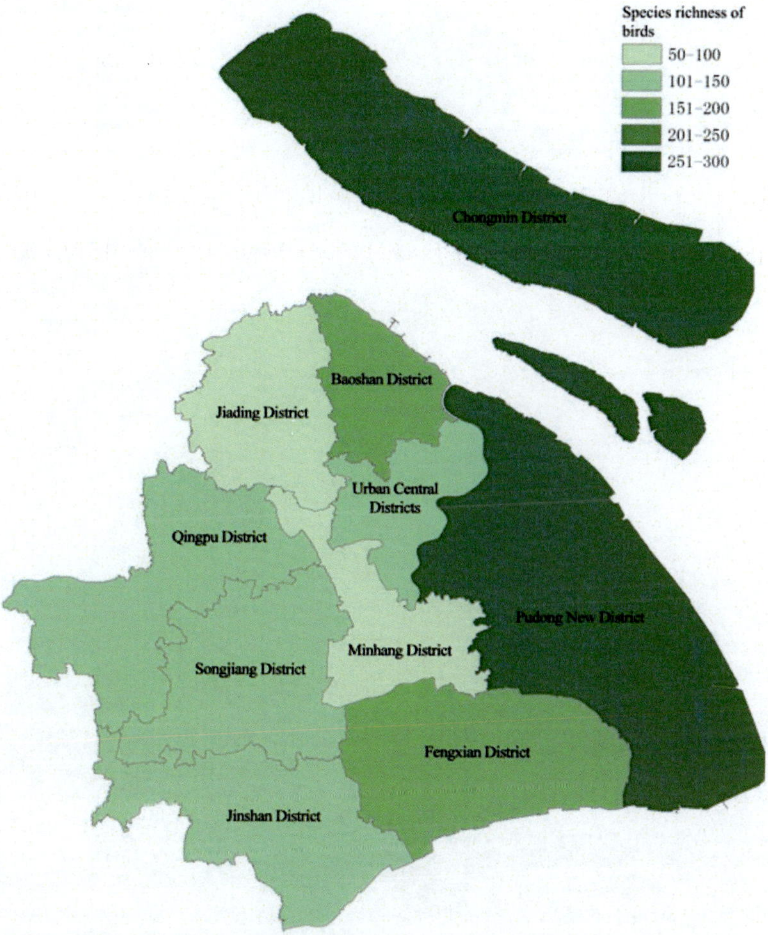

Fig. 3.5 Species richness of birds in each district of Shanghai (2011)

landscape graphs and calculated the number of river corridors and nodes for the statistical units of each district in the Shanghai ArcGIS map by using a spatial analysis tool. These calculations serve as the bases for validating the variables we used for the water network, including water network closure (α index), line–node rate (β index) and network connectivity (γ index).

α index: Network closure refers to the degree of loop for a landscape node in a connection network and is computed as $\alpha = (L–V + 1)/(2\ V–5)$, where L is the number of corridors, V is the number of nodes, $L–V + 1$ is the actual loop number and $2\ V–5$ is the maximum possible loop number.

β index: Line–node rate represents the average number of connections of each node in a water network and measures the complexity of a landscape node linked with other landscape nodes. This variable is computed as $\beta = 2L/V$.

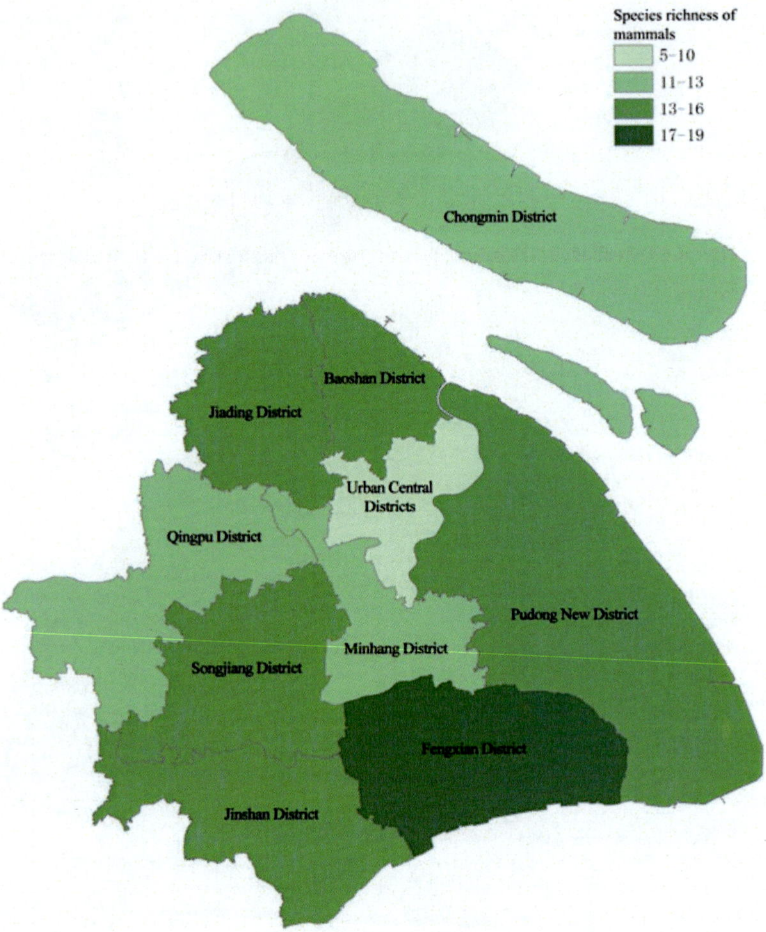

Fig. 3.6 Species richness of mammals in each district of Shanghai (2011)

γ index: Network connectivity describes the degree of connection amongst all landscape nodes in a single network and is computed as $\gamma = L/3(V-2)$.

(3) **Statistical analysis**

The IBM SPSS Statistics 19 software was used to carry out a Pearson's correlation analysis between the biodiversity variables and development intensity, ecological land and water network variables of each district unit. Two-sided tests for the statistical significance of the correlation were also conducted with P values of 0.01 and 0.05 to explore those main built environment influence factors that influence the biodiversity of Shanghai at the macroscale (Table 3.1).

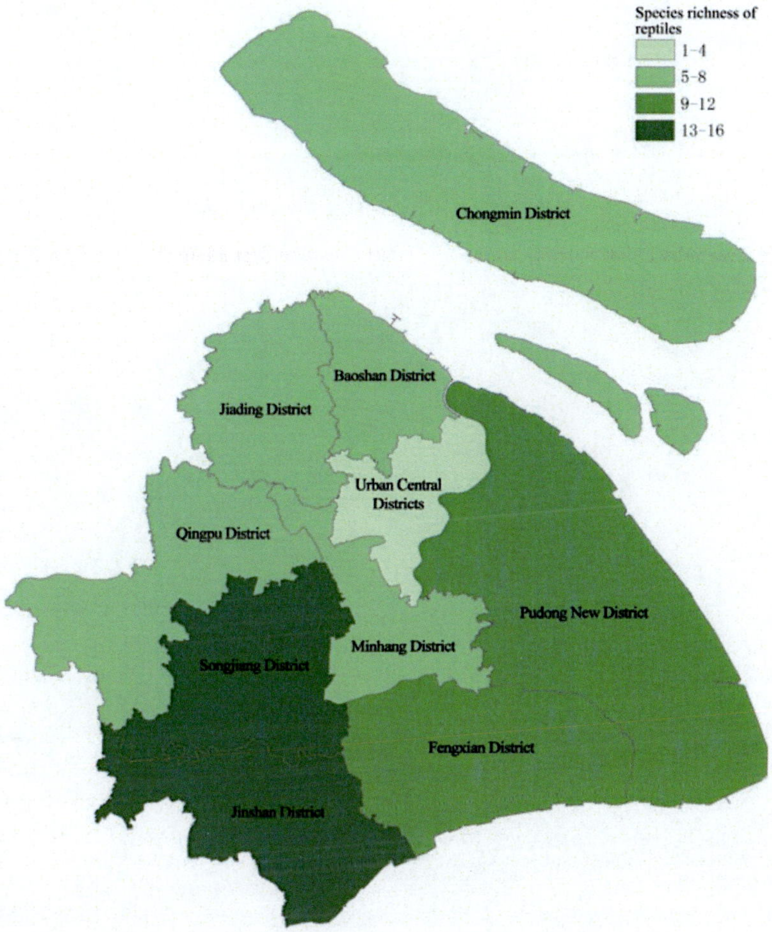

Fig. 3.7 Species richness of reptiles in each district of Shanghai (2011)

3.3 Correlation Analysis and Results

3.3.1 Correlation Analysis for Development Intensity and Biodiversity

The correlation analysis results are shown in Table 3.2. Population density, GDP per square meter and building density all show negative correlations with BI and species richness. Some variables show a significant negative correlation with a p-value of 0.01. The richness of invasive species shows a significant positive correlation with both population density and building density (p-value = 0.01), whereas GDP per capita is negatively correlated with BI and species richness of higher plants,

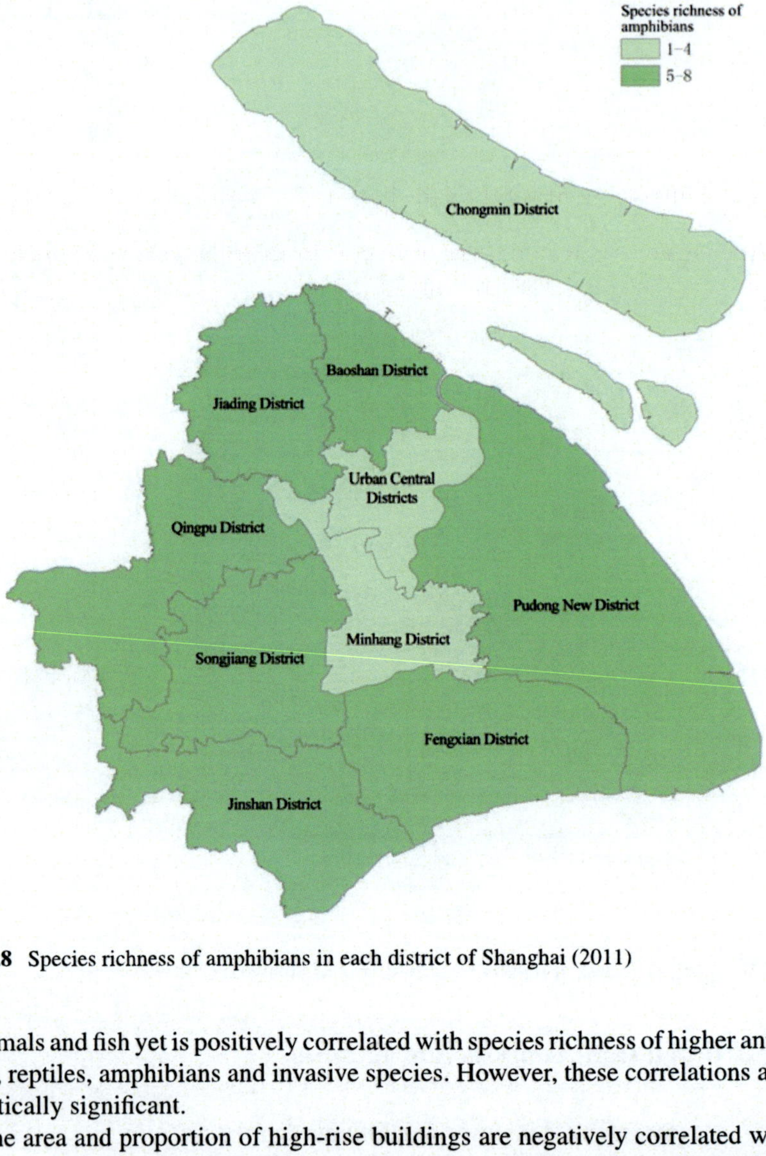

Fig. 3.8 Species richness of amphibians in each district of Shanghai (2011)

mammals and fish yet is positively correlated with species richness of higher animals, birds, reptiles, amphibians and invasive species. However, these correlations are not statistically significant.

The area and proportion of high-rise buildings are negatively correlated with BI and the species richness of each group yet are positively correlated with the richness of invasive species. When high-rise buildings with a different number of floors are considered, the proportion and area of those buildings with more than 16 floors are negatively correlated with biodiversity coefficient and species richness, especially in the case of animal groups. Meanwhile, the proportion and area of those buildings with more than 8 floors are negatively correlated with species richness, especially in the case of higher plant groups. Road network density shows a significant negative

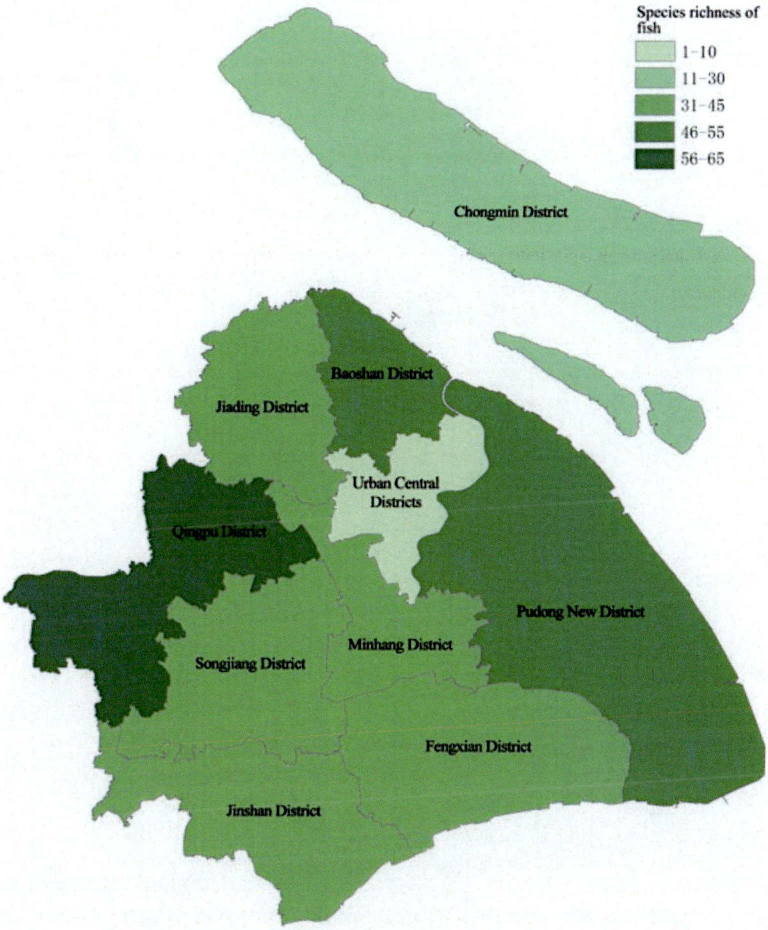

Fig. 3.9 Species richness of fish in each district of Shanghai (2011)

correlation with BI (p-value = 0.01) and species richness of fish (p-value = 0.05), a non-significant negative correlation with other species richness and a significant positive correlation with richness of invasive species (p-value = 0.01).

3.3.2 Correlation Analysis for Ecological Land and Biodiversity

As shown in Table 3.3, the ecological land area of Shanghai is significantly positively and negatively correlated with the species richness of birds (p-value = 0.01) and

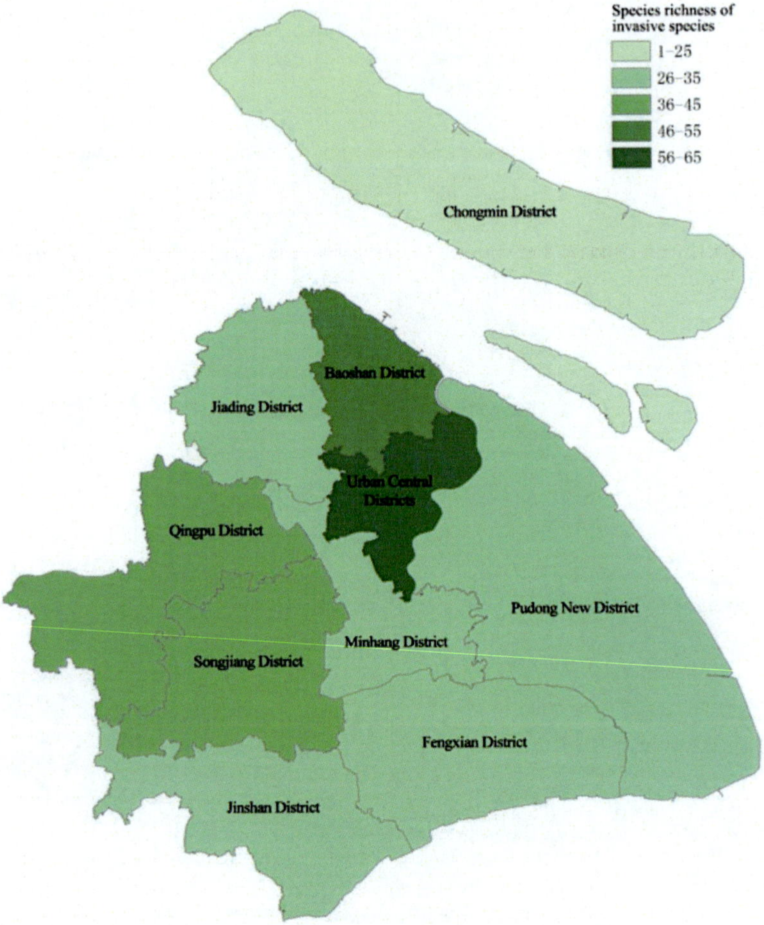

Fig. 3.10 Richness of invasive species in each district of Shanghai (2011). *Note* Data source of Fig. 3.2, 3.3, 3.4, 3.5, 3.6, 3.7, 3.8, 3.9 and 3.10 is Wang et al. (2012)

invasive species (p-value = 0.05), respectively. Major wildlife habitat area is significantly positive with species richness of higher animal (p-value = 0.01). Meanwhile, the proportion of ecological land area has a significant positive correlation with *BI* (p-value = 0.05) and a significant negative correlation with richness of invasive species (p-value = 0.05).

Amongst the four ecological land types analysed in Shanghai, urban waters and wetlands and cultivated land show the strongest influence on biodiversity. The area of urban waters and wetlands shows significant positive and negative correlations with the species richness of higher animals and birds (p-value = 0.05) and invasive species (p-value = 0.05), respectively, whereas their proportional area shows a significant positive correlation with the species richness of fish (p-value = 0.05).

Table 3.1 Urban built environment variables, their units and data sources

Type	Variable name	Unit	Data source
Development intensity	GDP per capita	10^4 RMB/person	Calculated using data from Shanghai Statistics Bureau (2012)
	GDP per square meter	10^8 RMB/km^2	
	Population density	person/km^2	
	Building density	10^4 m^2/km^2	
	High-rise building area	m^2	
	High-rise building area proportion	%	
	Road network density	km/km^2	Calculated using data from OpenStreetMap Foundation (2014)
Ecological land	Ecological land area	hm^2	Yang (2016)
	Urban public green space area	hm^2	
	Park area	hm^2	Shanghai Statistics Bureau (2012)
	Average area for a single park	hm^2	Calculated using data from Shanghai Statistics Bureau (2012)
	Cultivated land area	hm^2	Yang (2016)
	Urban water and wetland area	hm^2	
	Forest land area	hm^2	
	Major Wildlife habitat area	hm^2	Shanghai wildlife conservation management station (2008)
	Ecological land area proportion	%	Yang (2016)
	Urban public green space area proportion	%	
	Park area proportion	%	Calculated using data from Shanghai Statistics Bureau (2012)
	Cultivated land area proportion	%	Yang (2016)
	Urban water and wetland area proportion	%	
	Forest land area proportion	%	

(continued)

Table 3.1 (continued)

Type	Variable name		Unit	Data source
	Water network connectivity	α index	–	Calculated using data from Shanghai Urban Planning and Land Resource Administration Bureau (2013)
		β index	–	
		γ index	–	

A significant positive correlation is also observed between cultivated land area and species richness of higher animals (p-value = 0.05), whilst cultivated land area shows a significant negative correlation with the richness of invasive species (p-value = 0.05). The proportion of cultivated land area shows significant positive and negative correlations with *BI* and species richness of mammals (p-value = 0.05) and with the richness of invasive species (p-value = 0.01), respectively.

Biodiversity is positively influenced by the presence of forest land. The only significant negative correlation is found between the proportion of forest land area and the richness of invasive species (p-value = 0.05).

Urban public green space shows no obvious connections with *BI*, whereas its area has no significant correlations with any biodiversity variable. However, the large proportion of public green space area has a negative correlation with species richness (except for invasive species), a significant negative correlation with *BI* (p-value = 0.05) and a significant positive correlation with the richness of invasive species (p-value = 0.01). Meanwhile, the area of parks has a positive correlation with *BI* and the richness of the vast majority of the species as well as a significant positive correlation with the species richness of mammals and amphibians (p-value = 0.05). A positive correlation can also be observed from the average area of single parks to both *BI* and the species richness of each animal group. The average area of single parks shows a significant correlation with the species richness of higher animals and birds and a negative correlation with the richness of invasive species (p-value = 0.05). This finding indicates that in urban green spaces, those parks with a larger area have a stronger influence on biodiversity.

Table 3.3 also shows that α, β and γ indices of water network connectivity have a significant positive correlation with the species richness of higher animals (p-value = 0.05) and a positive correlation with the species richness of birds, amphibians, reptiles and fish. By contrast, these indices show a negative correlation with the species richness of invasive species and mammals. Therefore, urban water networks with better closure, more loops, better connectivity amongst nodes and network connectivity can have a positive effect on the richness of some species.

Table 3.2 Pearson's correlation analysis between development intensity and biodiversity at the district level in Shanghai

		Biodiversity index	Species richness							
			Higher plants	Higher animals	Of which					Invasive species
					Birds	Mammals	Reptiles	Amphibians	Fish	
Economic development level	GDP Per capita (10^4 RMB/person)	−0.258	−0.223	0.001	0.019	−0.374	0.007	0.062	−0.023	0.112
	GDP per square meter (100 million 10^8 RMB/km^2)	**−0.817****	−0.340	−0.294	−0.061	**−0.839****	−0.587	−0.582	**−0.692***	**0.761***
Population aggregation degree	Population density (persons/km^2)	**−0.827****	−0.344	−0.351	−0.125	**−0.827****	−0.613	−0.596	**−0.668***	**0.845****
Building intensity	Building density (10^4 m^2/km^2)	**−0.765****	−0.308	−0.325	−0.113	**−0.795****	−0.607	−0.533	−0.614	**0.907****
	Area of buildings with more than 8 floors (m^2)	**−0.639***	−0.413	0.007	0.177	**−0.724***	−0.384	−0.353	−0.480	0.475
	Area of buildings with more than 16 floors (m^2)	**−0.754***	−0.360	−0.154	0.069	**−0.819****	−0.529	−0.506	**−0.656***	**0.665***
	Area of buildings with more than 30 floors (m^2)	**−0.728***	−0.350	−0.111	0.110	**−0.806****	−0.512	−0.482	**−0.651***	**0.640***
	Area proportion of buildings with more than 8 floors (%)	−0.483	−0.410	−0.086	−0.003	−0.478	−0.127	−0.292	−0.231	0.158

(continued)

Table 3.2 (continued)

		Biodiversity index	Species richness							
			Higher plants	Higher animals	Of which					Invasive species
					Birds	Mammals	Reptiles	Amphibians	Fish	
	Area proportion of buildings with more than 16 floors (%)	**−0.699***	−0.377	−0.170	0.016	**−0.740***	−0.386	−0.445	−0.556	0.544
	Area proportion of buildings with more than 30 floors (%)	**−0.664***	−0.369	−0.002	0.195	**−0.752***	−0.437	−0.382	−0.572	0.550
	Road network density (km/km^2)	**−0.801****	−0.293	−0.468	−0.547	−0.499	−0.532	−0.283	**−0.738***	**0.829****

*Correlation is significant at a P value of 0.05 (two−sided test). **Correlation is significant at a P value of 0.01 (two-sided test)

Table 3.3 Pearson's correlation analysis between ecological land and biodiversity at the district level in Shanghai

	Biodiversity index	Species richness							
		Higher plants	Higher animals	Of which					Invasive species
				Birds	Mammals	Reptiles	Amphibians	Fish	
Ecological land area (hm^2)	0.419	−0.020	**0.780****	**0.772****	0.246	0.196	0.117	0.031	**−0.761***
Public green space area (hm^2)	−0.156	−0.334	0.313	0.308	−0.257	0.002	0.171	0.088	0.049
Park area (m^2)	0.226	−0.104	0.257	0.073	**0.701***	0.338	**0.730***	0.550	−0.395
Average area for a single park (hm^2/pcs)	0.427	0.022	**0.705***	**0.699***	0.280	0.208	0.074	0.008	**−0.757***
Cultivated area (hm^2)	0.464	−0.122	**0.701***	0.577	0.200	0.278	0.492	0.468	−0.596
Waters and wetland area (hm^2)	0.313	−0.028	**0.708***	**0.746***	0.187	0.057	−0.043	−0.113	**−0.660***
Forest land area (hm^2)	−0.077	0.026	0.095	0.073	−0.135	0.117	0.308	0.077	0.125
Major Wildlife habitat area (hm^2)	0.330	−0.363	**0.711***	0.597	−0.131	−0.101	0.438	0.482	−0.518
Ecological land area proportion (%)	**0.659***	0.297	0.487	0.398	0.557	0.405	0.244	0.228	**−0.763***
Public green space area proportion (%)	**−0.671***	−0.165	−0.549	−0.414	−0.581	−0.495	-0.368	−0.392	**0.903****
Cultivated land area proportion (%)	**0.637***	0.293	0.437	0.344	0.554	0.457	0.253	0.229	**−0.786****
Waters and wetland area proportion (%)	0.469	−0.076	0.277	0.076	0.209	0.187	0.571	**0.753***	−0.249

(continued)

Table 3.3 (continued)

		Biodiversity index	Species richness							Invasive species
			Higher plants	Higher animals	Of which					
					Birds	Mammals	Reptiles	Amphibians	Fish	
Forest land area proportion (%)		0.519	0.191	0.562	0.524	0.472	0.236	0.107	**0.086**	**−0.679***
Water network connectivity	α index	0.279	−0.316	**0.697***	0.620	−0.070	0.394	0.325	0.276	−0.395
	β index	0.317	−0.283	**0.693***	0.621	−0.146	0.377	0.282	0.260	−0.447
	γ index	0.264	−0.328	**0.696***	0.617	−0.053	0.393	0.335	0.281	−0.379

*Correlation is significant at a P value of 0.05 (two-sided test). **Correlation is significant at a P value of 0.01 (two-sided test)

3.4 Influence of Urban Built Environment on Biodiversity in Macroscale

3.4.1 *Influence of Development Intensity on Biodiversity*

The continued growth of human activities often has a significant negative impact on biodiversity (Dirzo and Raven 2003; Rands et al. 2010; Scanes 2018). The above results verify the significant influence of high-intensity urban development activities on local biodiversity, and the results show that population density, building density, high-rise building area and proportion and road network density are all significantly correlated with the majority of the biodiversity variables. A higher population density also corresponds to a larger building area, a stronger development intensity and a higher habitat fragmentation caused by traffic networks. Ortega–Alvareza and MacGregor–Fors (2009) and Glistra et al. (2009) obtained similar results. In economic terms, GDP per square meter shows a significant negative correlation with the majority of the biodiversity variables, thereby suggesting that economic output per unit area reduces the space for biodiversity to a certain extent. Similar to the findings of Wang et al. (2012), GDP per capita has a positive correlation with some biodiversity variables examined in this study; therefore, economic development does not fully entail a reduction in biodiversity. During the economic transition process of Shanghai, the development of a knowledge-intensive yet a non-labour-intensive industry may have positive effects on ecological environment protection.

The mechanism behind the influence of human development intensity in Shanghai on biodiversity can be interpreted as follows. Both population aggregation and urban high-density construction reduce the amount of ecological land spaces, such as urban waters and wetlands, cultivated lands and forest lands, thereby increasing the vulnerability of ecological systems and leading to loss of biodiversity. At the same time, whilst population aggregation intensity and traffic intensity increase, the exchange amongst species becomes more frequent, thereby increasing the occurrence of species invasion.

3.4.2 *Influence of Ecological Land on Biodiversity*

(1) **Water networks**

As ecological corridors, urban water networks provide diversified habitats for animals and plants. Energy flows, material flows and biological species often migrate and move along these networks (Paul and Meyer 2001; Rinaldo et al. 2018). Lynch et al. (2011) explored how the restructuring river connectivity changes may affect the patterns of freshwater fish biodiversity, but the influence of water network connectivity on other species remains unknown. Meanwhile, the results of this study reveal that a water network system with more network loops, better connectivity, more numbers of connections and stronger connectivity with water network landscape nodes has a positive effect on the species richness of birds, amphibians, reptiles and fish and has a negative effect on the species richness of mammals. In other words, those water networks with better connectivity can act as habitats and corridors for birds, amphibians, reptiles and fish and serve as a barrier for mammals. Given the limited number of wild mammals in metropolitan areas such as Shanghai, improving the landscape connectivity of water networks is beneficial for most species. Therefore, to enhance the role of rivers as corridor habitats that allow species to move and migrate smoothly and to improve the overall biodiversity, it is desirable to (1) increase the proportion and number of urban water network loops in a water network; (2) improve the loop connection system, (3) increase the number and scale of nodes in a water network and (4) enhance the accessibility of urban water networks and density of nodes.

(2) **Urban waters and wetlands**

The retention of water and wetland densities in human-dominated landscapes is essential in the conservation of aquatic and wetland-dependent organisms, such as fish and migratory birds (Gibbs 2000; Allen and O'Connor 2000; Reddy 2014). Shanghai is located in migratory routes between East Asia and Australia and serves as an important stopover and wintering habitat for a large number of water birds. Therefore, wetlands serve an important function in maintaining the biodiversity of Shanghai.

Amongst all types of ecological land areas, urban water and wetland areas show the most significant positive correlation with the species richness of birds, thereby suggesting that protecting a large number of key wetlands for birds and increasing the area of natural wetlands as much as possible are crucial in protecting avian diversity in Shanghai.

The proportion of urban water and wetland areas is related to the species richness of fish. Therefore, inland lake water systems play important roles in maintaining fish species. Urban water and wetland areas also show a significant correlation with the species richness of higher animals due to the fact that birds represent the largest proportion of higher animals in Shanghai. Therefore, any variable that is highly correlated with the species richness of birds (such as urban water and wetland areas) is also significantly correlated with the species richness of higher animals on the whole.

Along with social and economic development and urban expansion, the former city water courses, lakes and ponds have been replaced by roads, buildings and other structures. To guarantee the diversity of higher animal species in Shanghai, the proportion and area of reserve waters and wetlands should be considered.

(3) **Cultivated land**

As semi-natural ecosystems disturbed by human activities, cultivated lands serve a certain biodiversity-supporting function (Kathryn 1995; Benton et al. 2003). The cultivated land area of Shanghai shows a significant influence on the species richness of higher animals, whereas the proportion of cultivated land area has a significant influence on both *BI* and richness of alien invasive species and a positive influence on the richness of other single species.

Given that original and near-natural habitats are very rare in most parts of the Shanghai metropolitan area, cultivated land is considered the second most natural habitat (after wetlands) in Shanghai that can be inhabited by wild animals. Therefore, enhancing the ecological protection of cultivated lands presents an efficient way of maintaining and enhancing biodiversity.

(4) **Forest land**

Many studies reveal that urban forest cover and configuration play key roles in achieving high biodiversity (Berg et al. 1994; Berg 1997; Godefroid and Koedam 2003; Alvey 2006; Alberti et al. 2007; Canedoli et al. 2018). Meanwhile, in this study, the influence of forest land on most biodiversity variables is not significant. Only the proportion of forest land area shows a significant negative influence on the presence of invasive species probably due to the limited number of natural forest lands in the Shanghai metropolitan area. Other areas in Shanghai only have man-made forests, thereby explaining their limited biodiversity-supporting service functions. However, expanding the forest area, increasing their coverage rate and improving their ecosystem service benefits remain important measures for enhancing biodiversity. In this case, we need to protect the natural state of urban forest lands whilst simultaneously increasing their area and proportion.

(5) **Public green space**

Although urban green spaces are considered important havens for native plant and animal populations in many regions (Mörtberg and Wallentinus 2000; Gregory and Baillie 1998; Mason 2000; Gaston et al. 2005), the influence of public green spaces on most biodiversity variables is not significant and even negative. The mere increase in the proportion of public green spaces cannot effectively promote biodiversity at the macroscale and may even introduce invasive species in the area. However, as one type of public green space with a large area, urban parks are positively correlated with *BI* and the abundance of most species. These parks have a great influence on mammals and amphibians, whilst the average area of a single park has a large positive influence on many biodiversity variables due to the fact that other types of green spaces as defined in China's *Standard for classification of urban green space* (CJJ 85-2017) (Ministry of Housing and Urban–Rural Development of People's Republic of China 2017), including green buffers (i.e. green spaces for environmental protection along highways, railways and other infrastructures that demand protection for their surrounding areas), squares and affiliated green spaces, are greatly disturbed by human activities. These areas may provide regulating and cultural ecosystem services and may act as stepping stones for certain wildlife, such as birds, at the micro scale (Gan et al. 2019). Meanwhile, urban (larger) parks may act as shelter islands in the overall green space system where animals can find suitable habitats and resources.

Therefore, when developing public green spaces in high-density metropolitan areas with limited land resources, close attention should be given to the development of large parks.

(6) **Major wildlife habitats**

Major wildlife habitats are important spaces for wildlife survival and reproduction outside the nature reserves. As show in the above research results, the total area of major wildlife habitats in district level has a significant positive correlation with overall higher animal species richness, and its proportion has a significant positive correlation with the species richness of fish. The relationship between its proportion and the species richness of other species needs to be further investigated.

Wildlife habitat is also a hot research topic in biodiversity protection. The content of wildlife habitat protection was added in the Law of People's Republic of China on the Protection of Wildlife (revised version 2018), transferring the core of wildlife protection in China from species protection to habitat protection. In urban ecological spatial planning with biodiversity protection, major wildlife habitats are often included in the scope of spatial control. Therefore, a further analysis of the biodiversity quality and its influencing mechanism of major wildlife habitats in Shanghai will be carried out in Sect. 3.5.

3.5 Distribution and Quality of Wildlife Habitats in Shanghai and Its Environmental Influence Mechanism

3.5.1 Natural Distribution Characteristics of Wildlife Habitats

Birds account for the majority of wild vertebrates in Shanghai, which determines the main distribution characteristics of Shanghai's wildlife habitats. From the perspective of natural distribution, the Yangtze River estuary in the periphery of the city and the coastal wetlands on the north shore of Hangzhou Bay are the most major wildlife habitats in the city, and the most pristine areas in Shanghai, nurturing more than 70% of the city's wildlife species. The farmland and woodland habitats in the suburbs also provide habitats for a considerable number of wildlife, and the biodiversity of woodland continues to increase. Since 2000, Shanghai's urbanization has been advanced rapidly, and the city center has been constantly expanding, eroding part of the farmland habitats that were originally suitable for wildlife. But at the same time, significant achievements have been made in the construction of parks and green spaces in Shanghai, providing new habitats for wildlife (Zhang and Zhang 2015).

3.5.2 Biodiversity Quality of Major Wildlife Habitats and the Impact of Built Environment

(1) Evaluation of biodiversity quality

In 2008, the Shanghai Wildlife Conservation Management Station and related scientific research institutions conducted a biodiversity survey and environmental evaluation on 45 major wildlife habitats. The survey and evaluation involved 21 woodland habitats, 4 hill habitats, 10 lake and reservoir habitats, and 10 tidal flat wetland habitats, with a total area of about 1.9 million km^2, in 10 suburban districts, namely Baoshan, Qingpu, Minhang, Jiading, Songjiang, Jinshan, Fengxian, Pudong, and Nanhui (in 2009, one year after the survey, Nanhui District was merged into Pudong New Area), and Chongming. Among them, the Dianshan Lake habitat in Qingpu District is the largest, covering an area of 4,793.5 hm^2. The smallest habitats are the Chashan Mountain in Jinshan District and the Beigan Mountain in Qingpu District, with an area of only 6.5 hm^2 and 9.7 hm^2 respectively. According to the types of urban multiple habitats defined in Chap. 2 of this book, most of the 45 habitats surveyed are near-natural habitats on the non-development land, and can be regarded the areas bearing the largest number of wildlife and plant species in Shanghai.

The survey found a total of 201 species of birds, 15 species of terrestrial mammals, and 27 species of amphibians and reptiles, accounting for nearly half of the species in Shanghai, and contributing significantly to urban biodiversity. According to the survey, the wildlife habitats in Jiading District and Qingpu District boast the largest total number of species, followed by those in Minhang District, Chongming District, Pudong New Area, Fengxian District, and Songjiang District, and then followed by Jinshan District, Nanhui District and Baoshan District. The richness of wildlife habitats in Nanhui District and Baoshan District is relatively low (Shanghai Wildlife Conservation Management Station 2008). In terms of bird diversity index, Fengxian District and Nanhui District have the highest index scores, followed by Chongming District, Baoshan District, Jinshan District, Jiading District, Songjiang District, whereas Qingpu District, Minhang District, and Pudong New Area have the lowest scores (Fig. 3.11). The wildlife and plant habitats in Qingpu District have the largest number of mammalian species, while Chongming District has the largest number of reptile species. The numbers of amphibians in the 10 suburban districts are similar (Fig. 3.12).

First, calculate the Shannon-Weiner diversity index based on the number of species recorded and the number of individuals in each species:

$$H' = 3.3219 \times \left(\lg N - \frac{1}{N} \sum_{i=1}^{s} n_i \lg n_i \right)$$

Calculate the corresponding Pielou's evenness index based on the Shannon-Weiner diversity index:

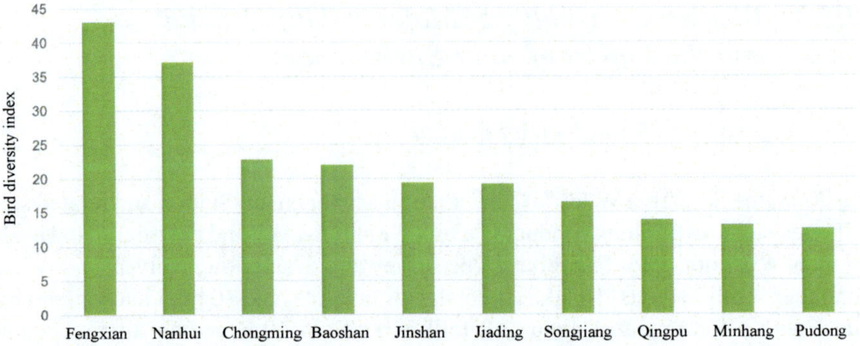

Fig. 3.11 Bird diversity index of major wildlife habitats in 10 suburban districts of Shanghai. *Note* The calculation method of the bird diversity index in the figure above is as follows

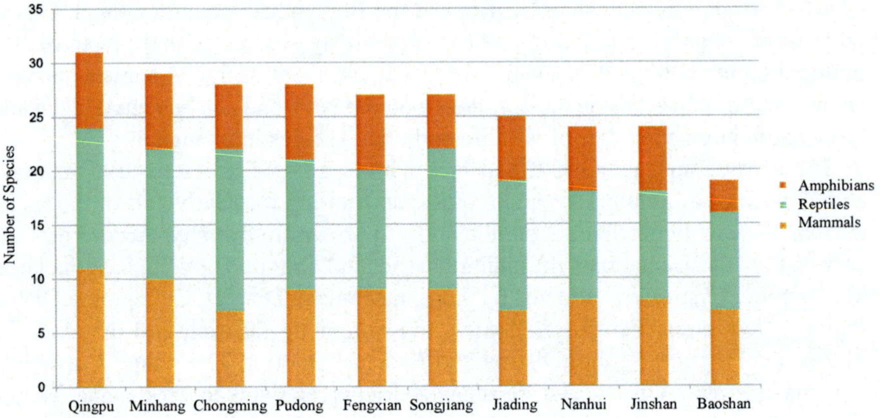

Fig. 3.12 Number of species of amphibians, reptiles and mammals in major wildlife habitats in 10 suburban districts of Shanghai

$$J = \frac{H}{H_{\max}} H_{\max} = \log_2 S$$

Among them, N is the total number of individuals in all species, ni is the number of individuals in the i species, and S is the total number of species.

For diversity index (D), $D = S \times J$, the definitions of parameters are the same as above.

Among the 45 major wildlife habitats surveyed, the forest for water resources conservation of the Qiqian River in Waigang Town, Jiading District, the forest for scenery on the Liuhe Island, the Malu Seedling Base, the forest for water resources conservation of the Suzhou Creek in Huacao Town, Minhang District, the Pujiang

Forest Park, the Heqing Gaohong Village in Pudong New Area, and the Dianshan Lake in Qingpu District are the habitats winning the highest scores for biodiversity. The Gucun Outer Ring Forest Belt in Baoshan District and Shanghai Weimeng Habitat Forest in Songjiang District are the two habitats with the lowest scores. On the whole, the coastal tidal flat wetland habitats in outer suburbs and the lake-reservoir habitats around the Dianshan Lake score higher when it comes to the indices of species richness and bird diversity, boasting better quality and higher conservation value, whereas the woodland habitats in suburbs score lower, which reduces their conservation value significantly (Figs. 3.13, 3.14, 3.15 and 3.16).

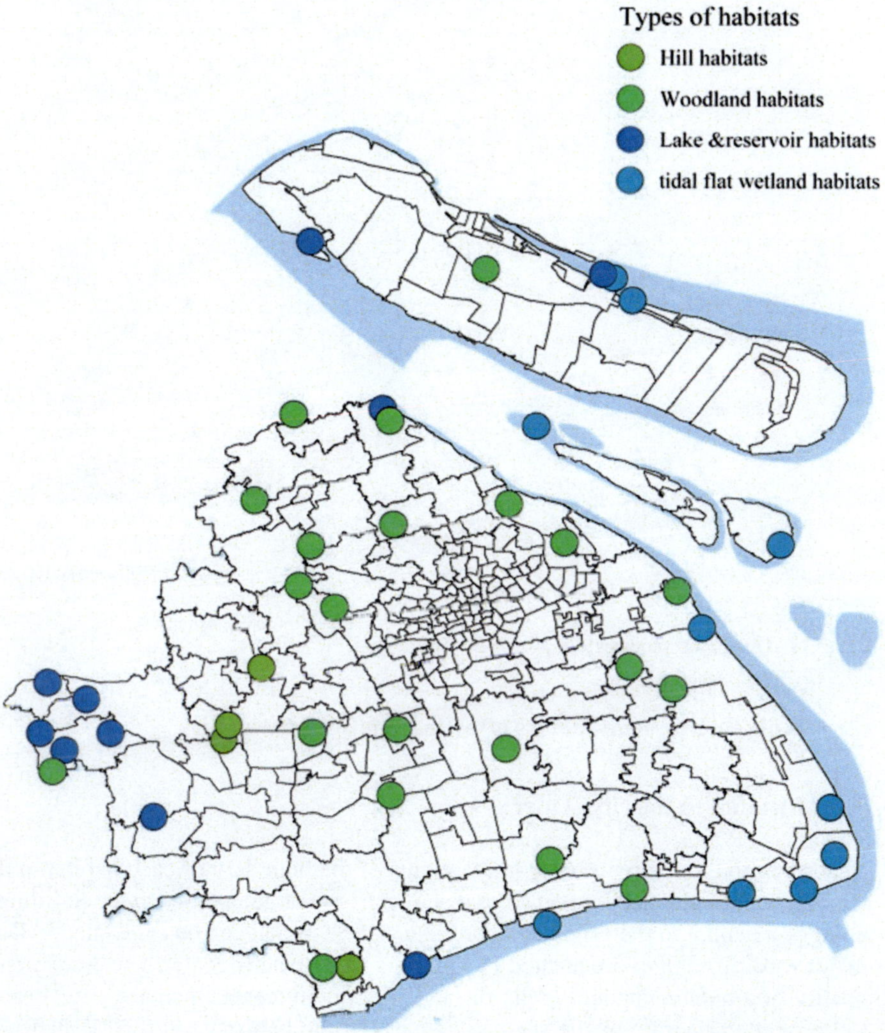

Fig. 3.13 Types of major wildlife habitats in Shanghai

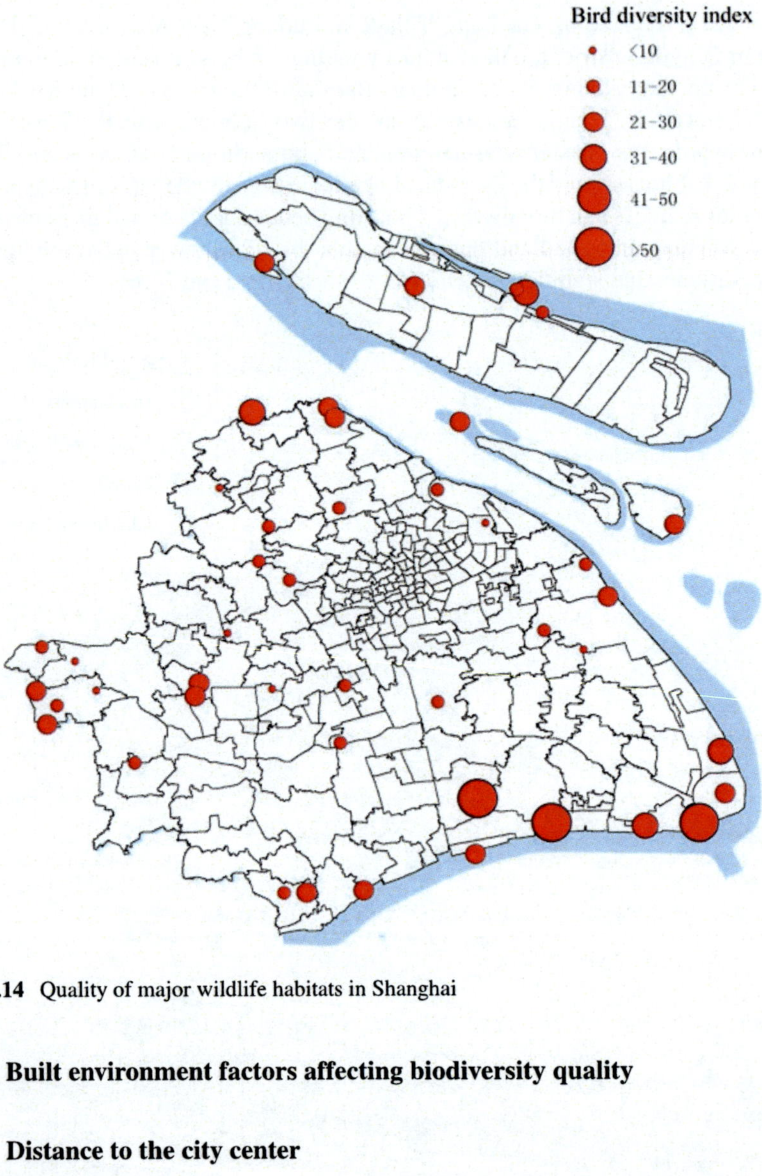

Fig. 3.14 Quality of major wildlife habitats in Shanghai

(2) Built environment factors affecting biodiversity quality

(1) Distance to the city center

According to the reference coordinate points of the boundary of each habitat in the "Survey and Evaluation Report," the distance from the geometric center coordinate point of a habitat to the origin of coordinates of Shanghai can be calculated. All 45 major wildlife habitats are located 13 km away from the city center. The biodiversity quality of a habitat changes as its distance to the city center increases. According to the scatter diagram of the species richness of the habitats and their distance to

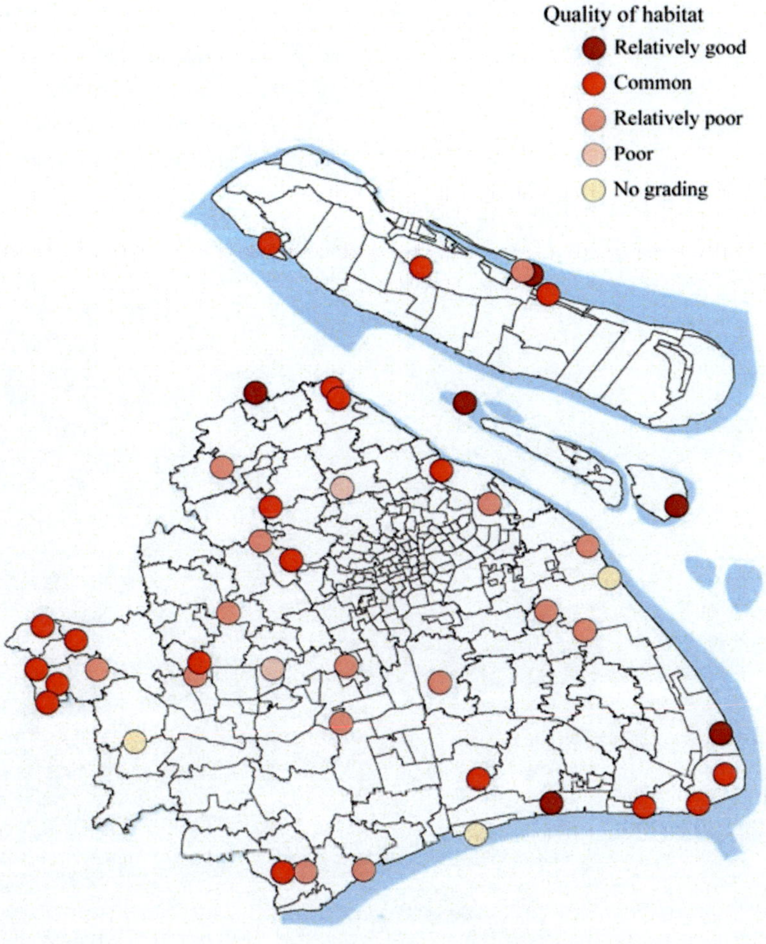

Fig. 3.15 Bird diversity index of major wildlife habitats in Shanghai

the city center (Fig. 3.17), these habitats are divided into three groups: 13.30 km, 30–45 km, and 45–60 km from the city center, namely the outer ring green belt to the peri-urban suburb green ring (the first circle), the peri-urban suburb green ring to outer suburb ring corridor (the second circle), and the outer suburb ring corridor to the metropolitan boundary (the third circle).

In the suburbs 13.30 km from the city center, there are 16 major wildlife habitats, of which the species richness ranges from 20 to 60, with an average value of 40; In the outer suburbs 30–45 km from the city center, there are only 9 major wildlife habitats, but they all have a relatively high species richness (40–110), with an average value of 65. It is worth mentioning that there are as many as 108 species of wildlife

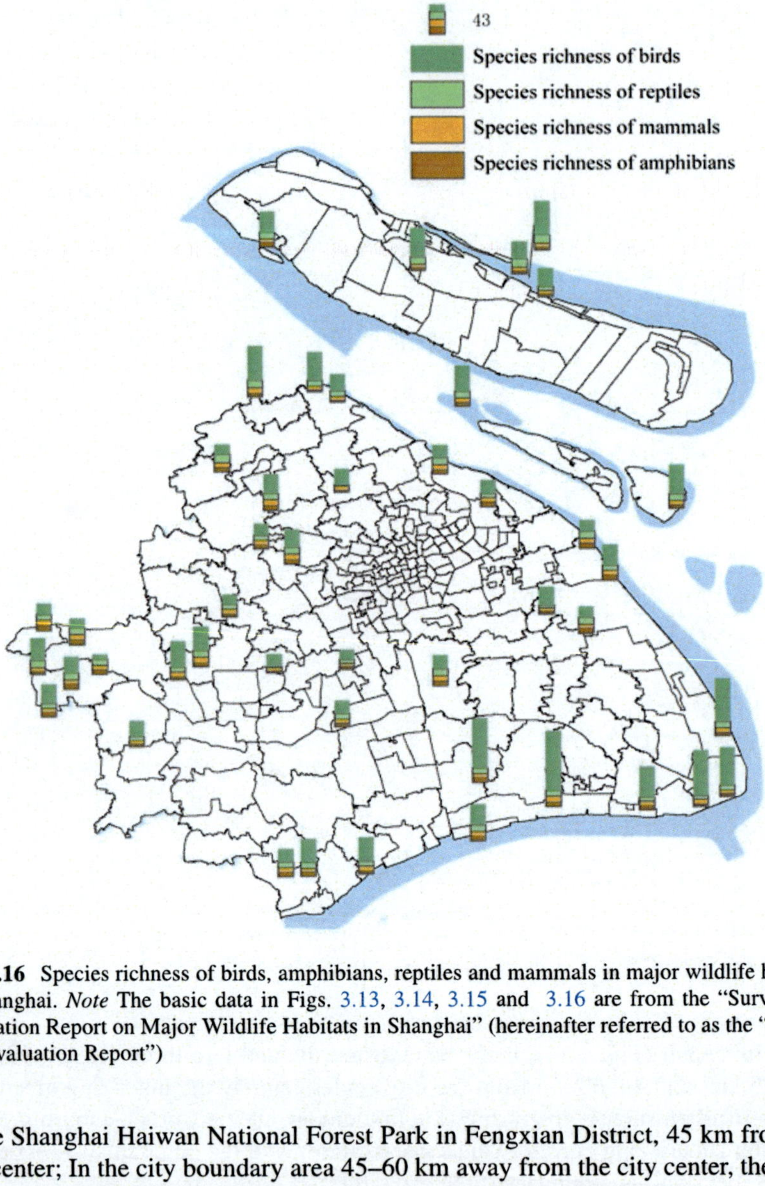

Fig. 3.16 Species richness of birds, amphibians, reptiles and mammals in major wildlife habitats in Shanghai. *Note* The basic data in Figs. 3.13, 3.14, 3.15 and 3.16 are from the "Survey and Evaluation Report on Major Wildlife Habitats in Shanghai" (hereinafter referred to as the "Survey and Evaluation Report")

in the Shanghai Haiwan National Forest Park in Fengxian District, 45 km from the city center; In the city boundary area 45–60 km away from the city center, there are 20 major wildlife habitats, most of which are coastal tidal flat wetland habitats and lake and reservoir habitats around the Dianshan Lake. The species richness is 30–80, with an average value of 52, which is slightly lower than that in the 30–45 km section (Fig. 3.17).

This pattern neither conforms to the theory that urban biodiversity increases along the urbanization gradient from the city center to the suburbs, nor does it completely conform to the intermediate disturbance hypothesis that the suburbs have the highest

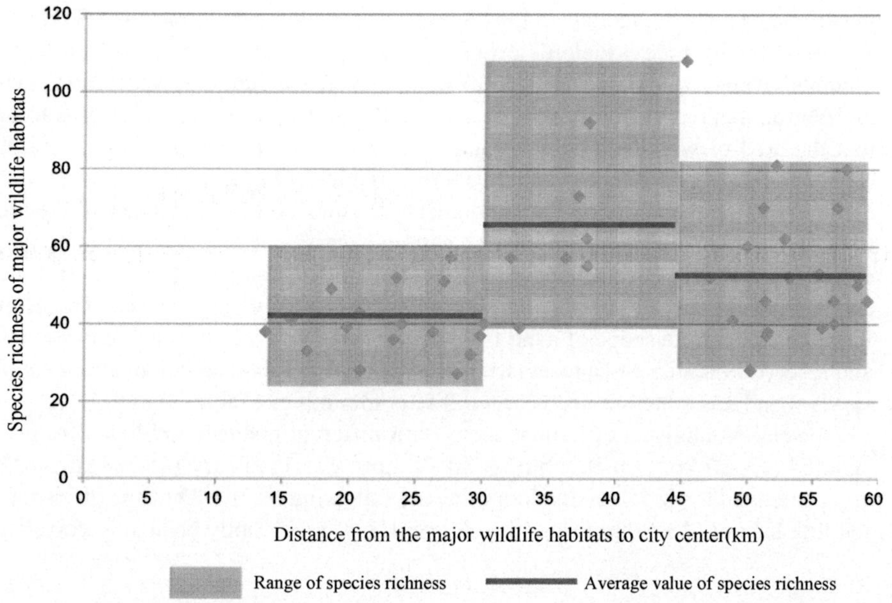

Fig. 3.17 Relationship between the species richness of major wildlife habitats in Shanghai and their distance to the city center

biodiversity. The reason behind may be the sprawling development of Shanghai. On the urbanization gradient, the peri-urban suburbs have been largely occupied by urban built-up areas. Although large in number, wildlife habitats in the peri-urban suburbs are besieged and poor in quality because of the close interference of urban development and human activities. However, the outer suburbs have become the intersection of urban built-up areas, rural areas and natural habitats, and have a better ecological environment. Although there are less important habitats, their species richness is higher. The ecosystem type of the coastal tidal flat wetland habitats and lake and reservoir habitats on the city boundary is quite simple, so the species richness is slightly lower than that in the outer suburbs. Therefore, compared with the existing empirical research cases, the "intermediate disturbance zone" has moved a circle outside in high-density Shanghai. In order to prevent the further expansion of the city from eroding major wildlife habitats, the master plan should first ensure the biological habitat spaces in the second circle and the third circle, and set a buffer zone outside the biological spaces that have been eroded and surrounded in the first circle.

(2) **Habitat area**

In this study, an overall and sub-category correlation analysis is conducted for the habitat area and the species richness of birds, mammals, amphibians, and reptiles in the 45 major wildlife habitats in the "Survey and Evaluation Report," the Shannon-Weiner diversity index, Pielou's evenness index, and diversity index of birds, the Shannon-Weiner diversity index and Pielou's evenness index of vegetation. (The calculation method of the above indexes is in the notes below Fig. 3.11.) It is found that the area of woodland habitats and lake and reservoir habitats is significantly related to the species richness and diversity variables of higher animals (except fish) and the birds in the habitats. The total area of various types of habitats and the area of tidal flat wetland habitats also have a certain positive correlation with the richness of various species. Given that 53% of these major habitats are less than 100 hm^2 and 33% are less than 50 hm^2, the overall area is already very limited. Therefore, when delimiting the scope of habitats, especially that of woodland habitats and lake and reservoir habitats, it is necessary to ensure considerable habitat area to meet the survival and reproduction needs of wildlife communities (Table 3.4).

Correlation analysis of habitat areas with different distance gradients from the suburbs (13–30 km), outer suburbs (30–45 km), city boundary (45–60 km) to the city center and the biodiversity data show the following results: The area of 16 major wildlife habitats 13–30 km from the city center is significantly positively correlated

Table 3.4 Pearson's correlation analysis of biodiversity and habitat area of major wildlife habitats in Shanghai

Variables			Area of all types of habitats	Type of habitats		
				Area of woodland habitats	Area of lake and reservoir habitats	Area of tidal flat wetland habitats
Species richness	Higher animals (except fish)		0.234	**0.824****	**0.676***	0.280
	Including	Birds	0.206	**0.853****	**0.721***	0.230
		Mammals	0.121	0.050	−0.030	0.048
		Reptiles	0.097	−0.002	0.067	0.219
		Amphibians	0.134	0.272	**−0.768***	−0.053
Bird diversity	Shannon–Weiner diversity index		0.044	0.385	0.268	−0.325
	Pielou's evenness index		−0.041	0.015	−0.264	−0.380
	diversity index		0.145	**0.779****	0.611	−0.093
Plant diversity	Shannon-Weiner diversity index		−0.117	−0.230	−0.144	0.622
	Pielou's evenness index		−0.216	−0.087	−0.248	0.482

Note 1. *Correlation is significant at a P value of 0.05 (two-sided test). **correlation is significant at a P value of 0.01 (two-sided test)

2. Dianshan Lake habitat mainly relies on the largest freshwater lake in Shanghai, and its area is not on the same quantitative level as other lake and reservoir habitats, so its data is excluded from the relevant analysis

3. There are only 4 hill habitats. Since the sample size is too small, no separate correlation analysis is conducted here

with the species richness of higher animals (except fish) and the birds in the habitats at a P value of 0.01, and significantly positively correlated with the diversity index and the Shannon-Weiner diversity index of the birds at a P value of 0.01 and 0.05, respectively. The area of 9 habitats 30–45 km from the city center is significantly positively correlated with the species richness of higher animals (except fish) and the birds in the habitats at a P value of 0.05. The area of 20 habitats 45–60 km from the city center has no significant correlation with their biodiversity. Therefore, when delineating major wildlife habitats in suburbs and outer suburbs, special attention should be paid to the area effect of habitats. Especially in the suburbs where the habitats have been gradually eroded by human development activities, it is necessary to avoid further erosion of the habitats by urban development (Table 3.5).

(3) **Disturbance from urban development around habitats**

The "Survey and Evaluation Report" divides the types of land in and around the habitat into 16 types: ① residential areas (including residences, primary and secondary schools, etc.); ② commercial areas; ③ cultural and educational areas (university campuses); ④ farmland; ⑤ industrial land (chemical industry); ⑥ industrial land (low pollution industry); ⑦ power stations; ⑧ tidal flats; ⑨ (near) natural wetlands (lakes, wetland parks, etc.); ⑩ reservoirs; ⑪ecological forests; ⑫nurseries; ⑬public green spaces; ⑭near-natural woodlands; ⑮parks; ⑯resorts. In this study,

Table 3.5 Pearson's correlation analysis of biodiversity and habitat area of major wildlife habitats in different areas of Shanghai

Variables			Habitat area 13–30 km from the city center	Habitat area 30–45 km from the city center	Habitat area 45–60 km from the city center
Species richness	Higher animals (except fish)		**0.655****	**0.705***	0.092
	Including	Birds	**0.787****	**0.734***	0.394
		Mammals	−0.077	−0.205	0.072
		Reptiles	0.157	−0.305	0.119
		Amphibians	0.149	0.384	0.146
Bird diversity	Shannon-Weiner diversity index		**0.532***	0.203	−0.262
	Pielou's evenness index		0.315	−0.253	−0.309
	diversity index		**0.775****	0.639	−0.044
Plant diversity	Shannon-Weiner diversity index		0.011	−0.647	0.015
	Pielou's evenness index		0.043	0.236	0.055

Note *Correlation is significant at a P value of 0.05 (two-sided test)
**Correlation is significant at a P value of 0.01 (two-sided test)

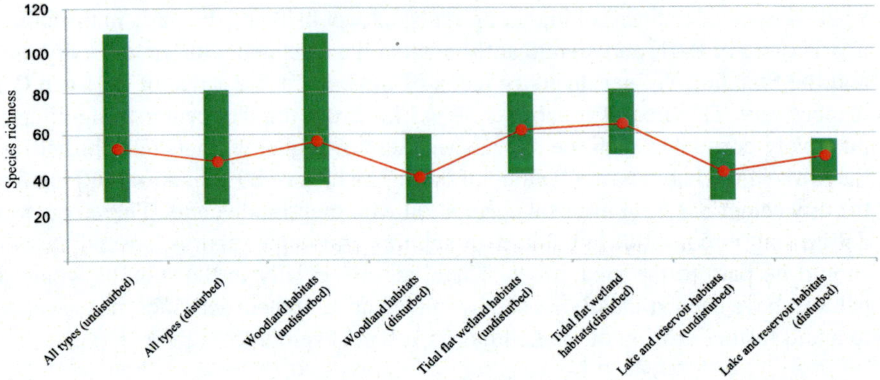

Fig. 3.18 Fluctuation range and average value of animal species richness in disturbed and undisturbed major habitats

the 6 types of land around habitats (①–③, ⑤–⑦) are classified as surrounding urban development activities. Such habitats are defined as those that are greatly disturbed by surrounding urban development activities, and other habitats are basically undisturbed ones (Fig. 3.18).

The results of the study show that: overall, the lowest and highest values of species richness in undisturbed habitats are higher than those in disturbed habitats, with a higher average value. This phenomenon also appears in woodland habitats, and the difference is more prominent. However, this phenomenon does not occur in tidal flat wetland habitats and lake and reservoir habitats, and the average species richness of disturbed and undisturbed habitats is almost equal. Even the average species richness in the disturbed habitats of the two categories is greater than that in the undisturbed habitats. This may be due to the fact that tidal flat wetland habitats and lake and reservoir habitats are relatively large in area, making them less susceptible to interference from surrounding development activities, while woodland habitats are relatively small, rendering the quality of their biodiversity more susceptible to surrounding development activities.

3.5.3 Analysis on the Spatial Control and Current Status of Major Wildlife Habitats

In terms of spatial control of urban wildlife habitats, Shanghai has established 4 nature reserves before 2008 (Dongtan Birds National Nature Reserve, Shanghai Jiuduansha Wetland National Nature Reserve, Jinshan Islands Marine Ecological Nature Reserve, Chinese Paddlefish Natural Reserve at the Yangtze River Estuary), 1 wildlife sanctuary (Nanhui Dongtan Wildlife Sanctuary), and 40 major habitats for wildlife across various districts, mainly woodlands and forest parks. A "4 + 1 + 40" protection network for wildlife and plants has gradually taken shape, with a total area of more than 1 billion m^2, accounting for 13.25% of the total area of Shanghai (Shen, 2011). In 2013, Fengxian District was included in the wildlife sanctuary, and nature reserves and major wildlife habitats were gradually included in the *Planning of Basic Ecological Network Structure in Shanghai* and other ecological space management and control plans for hierarchical protection and management.

The *Planning of Basic Ecological Network Structure in Shanghai* promulgated in 2012 puts forward the core goal of biodiversity conservation, through basic ecological space, suburban ecological space, ecological system around the city center, and green space system in concentrated urban areas to protect the ecological protection red line (Zhang 2013). The *Plan for Major Functional Zones of Shanghai* promulgated at the end of 2012 divides the land space of Shanghai into four types of functional areas: urban function optimization areas, new urban development areas, new urbanization areas, and comprehensive ecological development areas. Within the above four types of functional areas, there are pieces or dots of "development-restricted zones" and "development-prohibited zones" distributed based on their functional positioning, providing suitable space for wildlife to inhabit (Shanghai Municipal Government 2012). The *Planning Scheme for Drawing the Ecological Protection Red Line of Shanghai* released in 2015 draws ecological protection red line of the city to delimit 15 types of ecological spaces, including nature reserves, drinking water conservation areas, and major wildlife habitats, with an initially delineated total area of 4,364 km^2, proposing hierarchical project access requirements and implementation strategies for the identified primary and secondary protected areas (Shanghai Urban Planning and Land Resource Administration Bureau 2015).

However, rapid urbanization has been causing increasing fragmentation of the terrestrial wildlife habitats in Shanghai. We use Google Earth to visually inspect the status of 45 major wildlife habitats surveyed in 2008, and find that the habitats such as the Wuqi Farm in Luchao Port in Nanhui of Pudong New Area and the ecological forest in Gaohang Town have been eroded by urban development. After comparing with the *Planning of Basic Ecological Network Structure in Shanghai*, we find that these two habitats are not included in the ecological space network for management and control. The Google Earth survey results also show that the development of the two habitats occurred after 2010, and the large-scale development on the ecological forest site in Gaohang Town did not start until after 2013. Obviously, it is a result of the failure to carry out effective planning, protection and control in urban development (Fig. 3.19).

Major wildlife habitats are an effective supplement to the model of wildlife nature reserves. Section 3.4.2 of this book has verified its value to the overall biodiversity of an area (and its area is significantly positively correlated with the species richness of higher animals). Consequently, to include areas with important habitat values for wildlife in key ecological spaces in the suburbs of Shanghai into ecological

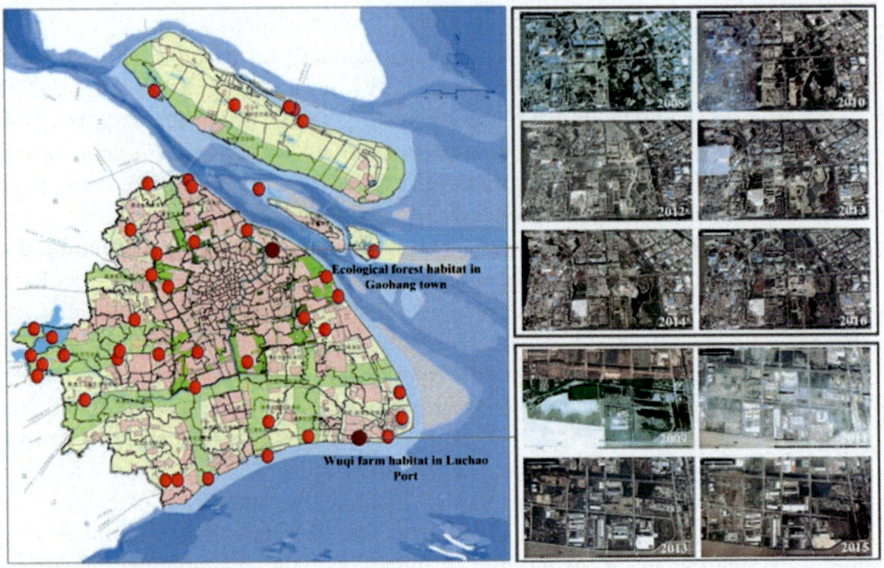

Fig. 3.19 Major wildlife habitats not included in the ecological space management and control have suffered erosion from urban development (Left: Overlay of 45 wildlife habitats and the "Planning of Basic Ecological Network Structure in Shanghai"; Right: The temporal and spatial evolution of the ecological forest habitat in Gaohang Town and the Wuqi Farm habitat in Luchao Port)

space management and control through statutory planning will protect their area and minimize the interference of surrounding urban development, further helping protect and strengthen regional biodiversity.

3.6 Summary

In this chapter, we take the district and county spatial units in Shanghai as an example to analyze the urban built environment variables that affect biodiversity on the macroscale, and focus on the distribution and quality of major wildlife habitats in Shanghai. The following four points have been discovered through survey.

(1) Negative influence of development intensity: Human development activities are the factors that have the greatest impact on biodiversity. Factors such as GDP per km^2, population density, building density, area of high-rise buildings, road network density and other factors are negatively correlated with the biodiversity index, of which the index values are all above −0.6 (strong negative correlation), and some are even as high as −0.8 (very strong negative correlation). Mammals and fish, which are more sensitive to environmental changes,

are more affected by development intensity than other species. Large-scale development has also brought in many invasive alien species. The essence of development control on the macroscale is to establish the boundary between development land and non-development land. Although non-development land resources are not necessarily used for biodiversity conservation, at least space is reserved for this purpose. Therefore, when considering the conservation and enhancement of biodiversity on the macroscale, we need to first consider imposing certain constraints on the proportion of the space occupied by human activities, to prevent the disorderly expansion of artificial development and reserve suitable space for biological habitation.

(2) Influence of the scale and proportion of ecological land: If the effect of development intensity is shown in determining the ratio of human and biological spaces, then the area of various ecological lands has a direct impact on the habitat quality for biological existence. For the city of Shanghai, the scale of waters and wetlands has the greatest impact on biodiversity, followed by the scale of cultivated land, so they need to be given priority in ecological space management and control. To ensure a stable population, whether it is the construction of parks and green spaces as semi-natural habitats or the construction of major habitats for wildlife as near-natural habitats, it is necessary to ensure the area effect of the habitats.

(3) Influence of water network pattern: Water network closure, line-node rate and network connectivity, all have a certain impact on the habitats and migration of wildlife. On the macroscale, to conserve and enhance biodiversity, in addition to reserving spaces for biological habitat conservation and ensuring sufficient habitat area, it is also necessary to consider the network channel effect between habitat nodes to ensure biological migration and the exchange of energy flow.

For details of the urban built environment variables that affect biodiversity and their influence effects on the macroscale, see Table 3.6.

(4) Ecological control of wildlife habitats: In the statutory urban planning, the incorporation of different levels of wildlife habitats into the scope of management and control is an effective means to ensure the space of multiple habitats for wildlife at the macro level. When delineating wildlife habitats, we should pay special attention to the remaining semi-natural habitats in the peri-urban areas to avoid shrinkage and interference from surrounding urban development.

Table 3.6 Urban built environment variables that affect biodiversity and their influence effects on the macroscale

Variable dimension		Variable type	Variable name		Influence effect
Habitat quality	Ecological land	Scale of land use	Area of ecological land (+)		Area effect
			Mainly include	Park area (+)	
				[Including the average area of individual parks (+)]	
				Area of cultivated land (+)	
				Area of major wildlife habitats (+)	
			Area proportion of cultivated land/water area and wetland/forest/major wildlife habitat (+)		Ratio effect
		Spatial form	Water network pattern (+)		Network effect
Human-induced disturbances	Development intensity	Level of economic development	GDP per km^2 (−)		Ratio effect
		Degree of population aggregation	Population density (−)		
		Development intensity	Building density (−)		
			High-rise building area ratio (−)		
			Road network density (−)		

Note "+" for positive effect, and "−" for negative effect

References

Alberti M, Booth D, Hill K et al (2007) The impact of urban patterns on aquatic ecosystems: an empirical analysis in Puget lowland sub-basins. Land Urban Plan 80(4):345–361

Allen AP, O'Connor RJ (2000) Interactive effects of land use and other factors on regional bird distributions. J Biogeogr 27:889–900

Alvey AA (2006) Promoting and preserving biodiversity in the urban forest. Urban Forest Urban Green 5(4):195–201

Benton TG, Vickery JA, Wilson JD (2003) Farmland biodiversity: is habitat heterogeneity the key? Trends Ecol Evol 18:182–188

Berg Å (1997) Diversity and abundance of birds in relation to forest fragmentation, habitat quality and heterogeneity. Bird Study 44(3):355–366

Berg A, Ehnstrom B, Gustafsson L et al (1994) Threatened plant, animal, and fungus species in Swedish forests: distribution and habitat associations. Conserv Biol 8:718–731

Canedoli C, Manenti R, Padoa-Schioppa E (2018) Birds biodiversity in urban and periurban forests: environmental determinants at determinants at local and landscape scales. Urban Ecosyst 21:779–793

Cantwell MD, Forman RTT (1993) Landscape graphs: ecological modeling with graph theory to detect configurations common to diverse landscapes. Landsc Ecol 8(4):239–255

Dirzo R, Raven PH (2003) Global state of biodiversity and loss. Annu Rev Environ Resour 28:137–167

Gan J, Guo GP, Yao XY (2019) Supporting services of urban roadside green spaces to biodiversity and their affections by built environment- case study of birds in roadside green space patches along century avenue, Shanghai. Landsc Archit 1:47–52 (in Chinese)

Gaston KJ, Warren PH, Thompson K, Smith RM (2005) Urban domestic gardens (IV): the extent of the resource and its associated features. Biodiv Conserv 14:3327–3349

Gibbs JP (2000) Wetland loss and biodiversity conservation. Conserv Biol 14:314–317

Glistra DJ, Devault TL, Dewoody JA et al (2009) A review of mitigation measures for reducing wildlife mortality on roadways. Landsc Urban Plan 91(1):1–7

Godefroid S, Koedam N (2003) How important are large vs small forest remnants for the conservation of the woodland flora in an urban context? Glob Ecol Biogeogr 12(4):287–298

Gregory RD, Baillie SR (1998) Large-scale habitat use of some declining British birds. J Appl Ecol 35:785–799

Kathryn F (1995) Assessing effects of agriculture on terrestrial wildlife: developing a hierarchical approach for the US EPA. Landsc Urban Plan 31(1–3):99–115

Lu L, He RJ, Liang J et al (2016) Research on basic ecological network planning strategy of Shanghai. Urban Construct Shanghai 2:62–68 (in Chinese)

Lynch HJ, Grant EHC, Muneepeerakul R (2011) How restructuring river connectivity changes freshwater fish biodiversity and biogeography. Water Resour Res 47:1–10

Mason CF (2000) Thrushes now largely restricted to the built environment in eastern England. Divers Distrib 6:189–194

Ministry of Housing and Urban-Rural Development of People's Republic of China (2017) Standard for classification of urban green space (CJJ 85-2017). China Architecture & Building Press, Beijing

Mörtberg U, Wallentinus HG (2000) Red-listed forest bird species in an urban environment assessment of green space corridors. Landsc Urban Plan 50:215–226

OpenStreetMap Foundation (2014) Street map of Shanghai. http://www.openstreetmap.org. Accessed 20 Jan 2018

Ortega-Alvareza R, MacGregor-Fors I (2009) Living in the big city: effects of urban land-use on bird community structure, diversity, and composition. Landsc Urban Plan 90(3–4):189–195

Paul MJ, Meyer JL (2001) Streams in the urban landscape. Annu Rev Ecol Syst 32:333–365

Rands MRW, Adams WM, Bennun L et al (2010) Biodiversity conservation: challenges beyond 2010. Science 329:1298–1303

Reddy PR (2014) Freshwater ecosystems and biodiversity—a case study of Kolleru Lake, India: a review. J Indian Geophys Union 18(2):277–288

Rinaldo A, Gatto M, Rodriguez-Iturbe I (2018) River networks as ecological corridors: a coherent ecohydrological perspective. Adv Water Resour 112:27–58

Scanes CG (2018) Human activity and habitat loss: destruction, fragmentation, and degradation. In: Colin Scanes CG, Toukhsati SR (eds) Animals and human society. Academic Press, London, San Diego, Cambridge, Kidlington, pp 451–482

Shanghai Environmental Protection Bureau, Shanghai greening and city appearance Administration, Shanghai Municipal Agriculture Commission, Shanghai Water Affairs Bureau (2013) Shanghai biodiversity conservation strategy and action plan (2012–2030). https://sthj.sh.gov.cn/fa/cms/upload/uploadFiles/2013-06-20/file1234.pdf. Accessed 20 Jan 2018

Shanghai Municipal Government (2012) Planning of Main Functional Areas in Shanghai (in Chinese)

Shanghai Statistics Bureau (2012) Shanghai Statistical Yearbook. http://www.stats-sh.gov.cn/html/sjfb/201701/1000198.html. Accessed 20 Jan 2018. (in Chinese)

Shanghai Urban Planning and Land Resource Administration Bureau (2013) Shanghai ArcGIS map. Shanghai Urban Planning and Land Resource Administration Bureau, Shanghai

Shanghai Urban Planning and Land Resource Administration Bureau (2018) Shanghai Master Plan (2017–2035). http://www.shanghai.gov.cn/newshanghai/xxgkfj/2035004.pdf. Accessed 20 Jan 2018. (in Chinese)

Shanghai Urban Planning and Land Resource Administration Bureau (2018) Atlas of Shanghai Master Plan (2017–2035). Shanghai Urban Planning and Land Resource Administration Bureau, Shanghai (in Chinese)

Shanghai Urban Planning and Land Resource Administration Bureau (2015) Planning scheme of Ecological Protection Redline in Shanghai (in Chinese)

Shanghai wildlife conservation management station(2008) Investigation and assessment report of important wildlife habitats in Shanghai, Shanghai wildlife conservation management station, Shanghai (in Chinese)

Shen ML (2011) Four nature reserves have been established in Shanghai, and four conservation laws have been issued to create a paradise for wildlife. Xinmin Evening, February 16 (in Chinese)

Wang Q, Ruan JJ, Sha CY et al (2012) Human impacts on the spatial pattern of biodiversity of Shanghai City. Ecol Environ Sci 21(2):279–285 (in Chinese)

Yang L (2016) Urban Ecological Land Optimization Study-Case Study of Shanghai Pudong New Area as example. Master's thesis, Tongji University, Shanghai (in Chinese)

Zhang YX (2013) Ecological spatial planning in light of rapid urbanization: innovation and implementation. Shanghai Urban Plan Rev 5:7–10 (in Chinese)

Zhang ZT, Zhang ED (2015) Urban wildlife habitat protection model: a case study of Shanghai. Chin J Wildlife 36(4):447–452 (in Chinese)

Chapter 4
Relationship Between Diversity and Built Environment on Meso-Micro Scale: The Case Study in Blocks Along Century Avenue, Pudong New District, Shanghai

4.1 Study Area and Methods

4.1.1 Birds as Studied Species

(1) Birds are the indicator species at the top of urban food chain

Wild birds in a city are an important part of urban wildlife. Birds are at the high or top trophic level of the food chain in the urban ecosystem, and are more sensitive to the composition of habitats and the accumulative effect of environmental pollution. An environment suitable for the existence of wild birds means that natural food sources such as insects, fruits, and nectar are relatively abundant, and that it has a good vegetation environment. Meanwhile, the special physiological structure of birds enables those fly long distances and make extensive use of various urban habitats of different scales to meet their physiological needs of foraging, resting, courtship, and nesting. Therefore, as a cosmopolitan species that adapts to a wide range of habitat types and a typical species with a broad-spectrum food sources in urban environment, birds are usually regarded as an indicator organism for the urban ecological environment, and play an irreplaceable indicating role especially in the study of small-scale, high-density and fragmented urban artificial habitats.

(2) Birds are easier to monitor in urban environment

The protection and appropriate use of wild bird resources has become one of the indicators of a country's natural environment, science, culture, and social civilization in the world. Birds are easy to be noticed by the public visually and acoustically, and are generally recorded in a more detailed way in the basic data of regional biological community. Moreover, the number and richness of their species are relatively easy to monitor. Therefore, bird communities and their species diversity are often used as important indicators of modern urban ecological health in urban ecological research.

J. Gan, *Urban Biodiversity and Built Environment*,
https://doi.org/10.1007/978-981-16-0949-7_4

(3) **Birds boast the largest number of species among wild vertebrates in Shanghai**

As Shanghai is an important transfer stop on the migratory birds' migration route from East Asia to Australasia, it is abundant in bird resources, and the number of bird species ranks first in the Yangtze River Delta region (Fig. 4.1). Birds are the most important wildlife group in Shanghai, accounting for more than 80% of the total number of wild vertebrate species in Shanghai (Zhang and Zhang 2015). From the beginning of the 20th century to 2013, the number of bird species recorded in the Shanghai area has reached 445 species of 20 orders, representing 70 families. The 2013–2015 "Second Survey of Terrestrial Wildlife Resources in Shanghai" (hereinafter referred to as the "Second Survey") recorded a total of 227,619 birds of 219 species. Based on this data, it can be estimated that there are more than 4.6 million birds of various species inhabited in Shanghai every year. Among them, migratory birds account for about 80%, including passing migratory birds (37%), winter migratory birds (29%) and summer migratory birds (14%), while resident birds and vagrant birds account for 14% and 8% respectively (Jin 2016). The water bird survey conducted at the same time as the "Second Survey" recorded 47,262 water birds of 52 species, including 601 birds of 6 species protected at national level, 2,267 birds of 5 species of the threatened species listed by the International Union for Conservation of Nature (IUCN), and a newly discovered water bird species in Shanghai. Specifically, Hengsha Dongtan and Nanhui Dongtan have the largest number of water birds that both exceed 10,000, followed by Chongming Dongtan and Jiuduansha, each with more than 7,000 (Shanghai Forestry Bureau 2016).

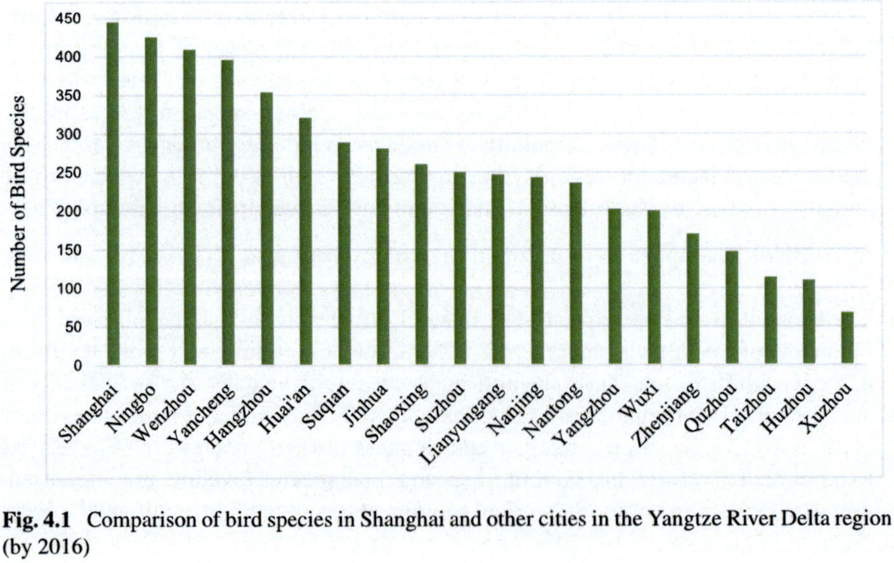

Fig. 4.1 Comparison of bird species in Shanghai and other cities in the Yangtze River Delta region (by 2016)

Data Sources:

Shanghai: Shanghai completed the water birds survey conducted at the same time as the second national survey of terrestrial wildlife resources. http://www.shanghai.gov.cn/nw2/nw2314/nw2315/nw4411/u21aw1097711.html;

Nanjing: Nanjing releases the status quo of biodiversity conservation. http://www.jiangsu.gov.cn/gzdt/201505/t20150523_386811.html;

Wuxi: Preliminary list of wildlife resources in Wuxi revealed, and 15 species of birds added in two years. http://news.jnwb.net/2013/1115/73212.shtml;

Xuzhou: Tian YY, Zhou H (2013) Changes in bird populations in Xuzhou in the first decade of the 21st century. Journal of Anhui Agricultural Sciences 36:13911–13913;

Suzhou: 10 species of migratory birds first seen in Suzhou, and a total of 254 species observed this year. http://js.people.com.cn/n2/2016/0601/c360304-28438726.html;

Nantong and Lianyungang: Wu JF et al. (2012) Composition and analysis of bird communities in the coastal areas of Nantong and Lianyungang. Hubei Agricultural Sciences 51(22): 5126–5130;

Huai'an: There are 321 species of birds in Huai'an. http://news.hynews.net/ha/2015-04-21/93362.html;

Yancheng: People of Yancheng have a strong awareness of loving and protecting birds, and the bird rescue team grows. http://js.ifeng.com/yc/news/detail_2015_09/15/4350124_0.shtml;

Yangzhou: How many species of wildlife and plants are there in Yangzhou? Birds are the "protagonists" among all animals. http://www.yznews.com.cn/news/2015-03/03/content_5478194.htm;

Zhenjiang: Wang HZ, Chen JB (2014) Current situation and countermeasures of wildlife protection and management in Zhenjiang City. Journal of Green Science and Technology 8: 62–63;

Taizhou: Wang J et al. (2015) Investigation of bird diversity in Taizhou area [J]. Journal of Nanjing Normal University (Natural Science Edition) 2: 116–121;

Suqian: More and more wildlife in Suqian in spring! http://www.suqian.gov.cn/sqlyjwz/snlhxwdt/201503/fea37d04eb644af49cdf602430cd36dc.shtml;

Hangzhou: Ding P et al. (2008) Resources of terrestrial wildlife in Hangzhou City. Journal of Chinese Urban Forestry 6(4):62–71;

Ningbo: Ningbo has recorded 425 species of birds. Where can citizens go to observe them? http://zj.zjol.com.cn/news/316918.html;

Wenzhou: Lin Guanle et al. (2014) Utilization status and management countermeasures of terrestrial wildlife resources in Wenzhou. East China Forest Management 4:31–33;

Huzhou: Dong L (2016) More than 200 species of wildlife and 1,334 species of wild plants: How much do you know about the wildlife and plant resources in Huzhou?. Huzhou Evening News, April 15;

Shaoxing: Shaoxing Municipal Forestry Bureau (2012) Survey report on bird resources in Shaoxing City;

Jinhua: Hu MX (2015) Let's protect wild birds together. https://www.jhnews.com.cn/;

Quzhou: Xiong S et al. (2014) Status and countermeasures of domestication, breeding, management and utilization of wildlife in Quzhou. East China Forest Management 3: 15–17;

No data were found in Changzhou, Jiaxing, Zhoushan, Taizhou, and Lishui.

(4) **Shortcomings in existing bird research**

Existing research mainly focuses on near-natural and semi-natural habitats, and there is a lack of research on the avian diversity in semi-artificial habitats in the attached green spaces where urban residents are most exposed to. There are about 60 species of common birds in various districts in Shanghai, mainly distributed in urban park and green space habitats, wetland habitats, farmland habitats and habitat on low mountains and hills in the metropolitan area of Shanghai (Zheng 2011). Urban birds and their habitats have always been the focus of Shanghai urban biodiversity research. Luan (2003) made an in-depth comparative study on bird communities and their habitats in four habitats: urban wetlands, woodlands, farmlands, and public green spaces. Tang et al. (2003), Ge et al. (2005), Lu et al. (2007), Yuan et al. (2011), and Yang et al. (2015) conducted surveys on wild birds in the following places including Changfeng Park, Longhua Park, People's Park, Daning Lingshi Park, Zhongshan Park, Yanzhong Greenland, Shanghai Botanical Garden, Century Park, Outer Ring Greenland, Jing'an Park, Shanghai Zoo, Pudong Gaoqiao Riverside Forest Park, and used statistical methods such as cluster analysis, regression analysis, and correlation analysis to analyze the environmental factors affecting the bird community structure and avian diversity. However, related researches mostly focus on near-natural and semi-natural habitats with relatively large areas, and few studies explore the impact of bird diversity and built environment in semi-artificial habitats in the attached green spaces in typical high-density urban areas such as urban residential areas, business areas, and areas for science, education and culture.

4.1.2 Selection of Study Area

This chapter selects the area along the Century Avenue in Pudong New Area, Shanghai as the study area. Starting from the Oriental Pearl Tower in the west and ending at the Century Park in the east, the Century Avenue is an axis connecting the city's two landmark nodes. The study area covers the Century Avenue from the little Lujiazui area to the area 100 m away from both sides of Yanggao Road and the complete neighborhood, and the neighborhood along the Jinxiu Road opposite the Century Park, a total of 54 blocks. As 3 blocks are under construction, 51 blocks were actually surveyed (Fig. 4.2). The study area includes Shanghai Lujiazui Financial Center, New Shanghai Commercial City, Zhuyuan Business District, Huamu Administrative and Cultural Center in Pudong New Area, together with multi-storey

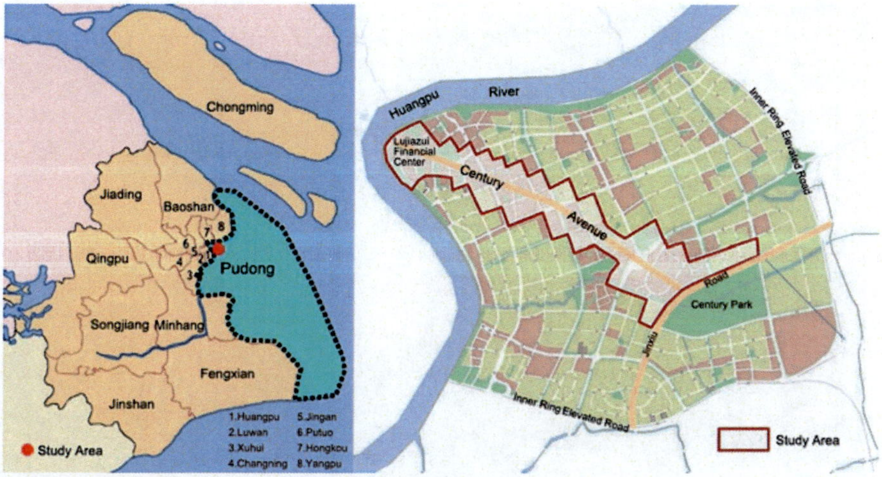

Fig. 4.2 Location map of study area on the meso-micro scale

commercial and residential areas represented by Dongchang Community, Dongyuan Village No. 1, Laoshan Villages No. 3 and 4, Rushan Villages No. 2 and 3, and the high-rise residential areas such as Lujiazui Central Apartment, Lianyangxinyuan Community, Tian'an Garden, and Hualijiazu Garden along the Jinxiu Road, with a total area of 4.60 km^2, of which the total development land area is 4.57 km^2 and the water area is 0.3 km^2 (Fig. 4.3).

The reasons for choosing this area: the area is located in the central area of the city, with convenient transportation, which facilitates research and observation. **The land-use types in this area are diverse**. According to the *Shanghai Land Use Classification Standard (2011)*, the land in this area are mainly divided into residential land (20.22%), land for public facilities (27.45%), land for roads and squares (29.51%), green spaces (14.79%), reserved land for public facilities (1.47%), etc. Specifically, the land for public facilities includes land for administration, land for commercial services, land for cultural facilities, land for business offices, land for education and research, etc., which shows the land structure of a typical urban center (Figs. 4.4 and 4.5).This area is a great specimen for studying the impact mechanisms of different land-use types and their built environment spatial patterns on biodiversity.

(1) **Diverse habitat types**

The habitats of the study area are mainly semi-artificial habitats in the attached green spaces composed of concentrated green spaces or green spaces between houses in residential areas, business areas, and commercial areas, near-natural habitats such as floodplain by the Huangpu River and Zhangjiabang waters, semi-natural habitats in the urban parks and public green spaces such as the Century Park, Lujiazui Central Greenland, Douxiang Park, and Oriental Pearl Park, and even a wetland habitat next to the parking lot of the Shanghai Science and Technology Museum, which is one of

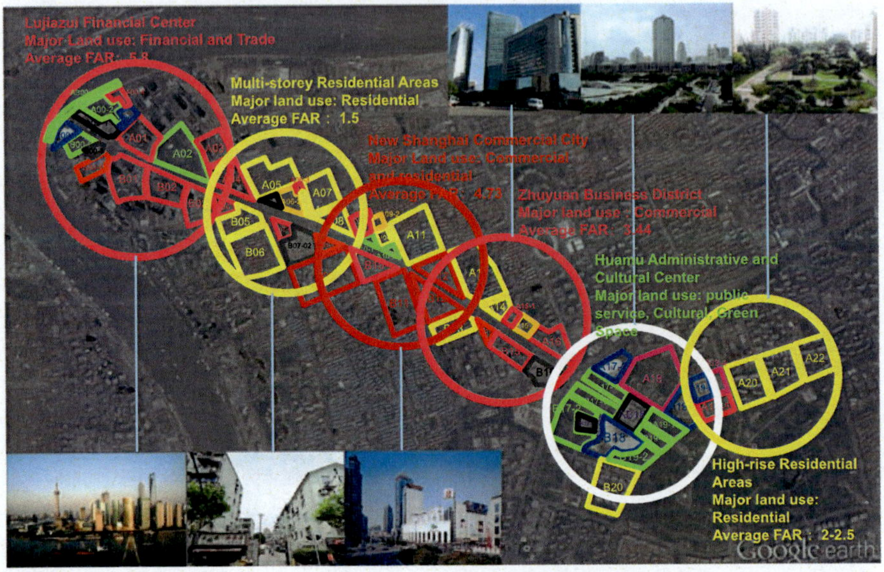

Fig. 4.3 Study area division and its development intensity index on the meso-micro scale

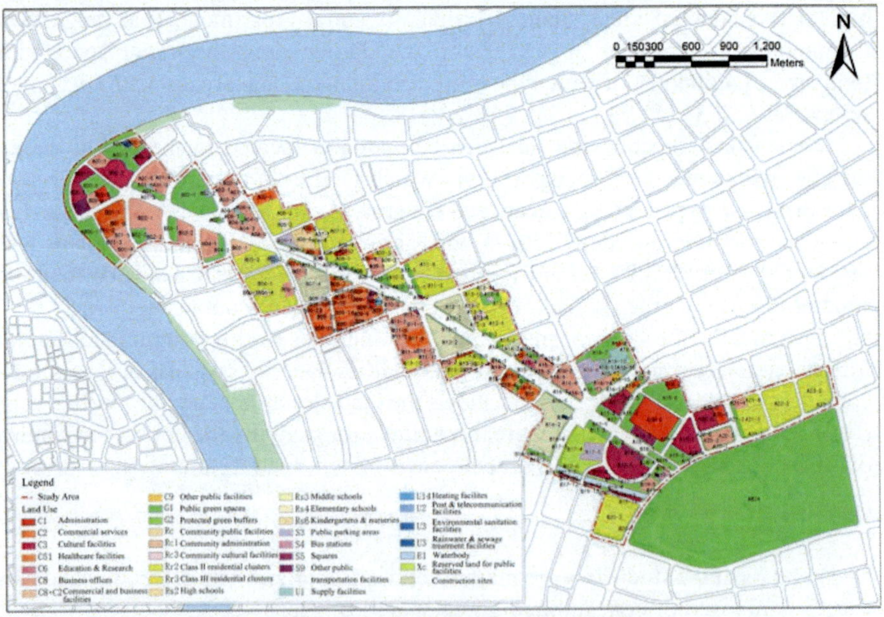

Fig. 4.4 Current status of land use in study area on the meso-micro scale

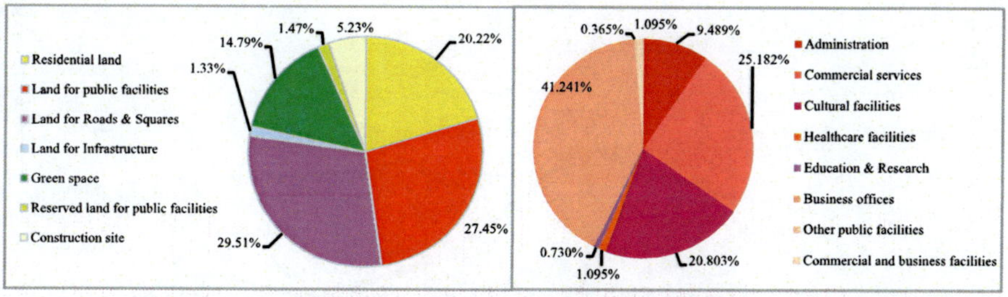

Fig. 4.5 Proportion of land use in study area on the meso-micro scale

the reserved land for public facilities of Huamu Administrative and Cultural Center. It has not been put into service in time after the completion of the Shanghai Science and Technology Museum and the Shanghai Oriental Art Center. After more than 10 years of natural succession, a rare natural wetland has taken shape here in the central area of Shanghai. In addition, because the Century Avenue cuts into the existing urban spatial texture, a series of triangular temporary roadside green spaces of varying sizes have taken shape on both its sides. These diverse habitats provide various birds with places for foraging and habitation, and also become a potential space for bird activities in the entire study area (Fig. 4.6).

Fig. 4.6 Various habitat types in study area on the meso-micro scale

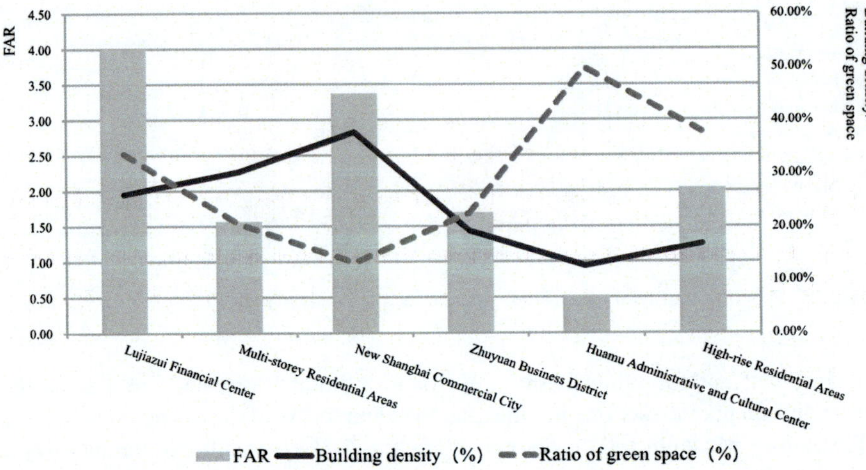

Fig. 4.7 Gradient of FAR, building density and Green space ratio in the six large study areas on the meso-micro scale

(2) **Continuous urban blocks**

Existing similar studies mostly select several different types of land sections in different location gradients of the city for investigation and analysis, and the results of those studies are extremely susceptible to interference from different surrounding conditions of the sample blocks. In this study, continuous urban blocks are used as sample blocks to present the continuous distribution pattern of bird biodiversity in a more objective way. In addition, the six large areas involved in the sample area show obvious density gradient levels of urban spatial agglomeration (Fig. 4.7), which also helps to measure the continuous impact of different density gradient sections on biodiversity.

4.1.3 Survey Methods

(1) **Bird surveys**

By referring to the bird survey methods proposed in the *Technical Guidelines for Biodiversity Monitoring—Birds* (Ministry of Environmental Protection 2014) issued by the Ministry of Environmental Protection in 2014, we divided the study area into 5 survey zones, each with 10–25 fixed survey sampling points, and one line transect covering the main habitats, landscapes and vegetation types in the block (Fig. 4.8). Survey time was from November 2014 to October 2015, and the surveys were conducted in the first ten days of each month. The surveys were mainly conducted in

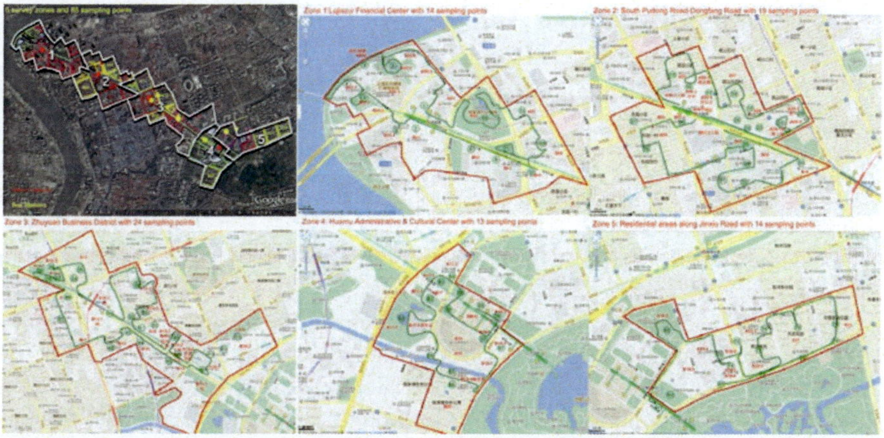

Fig. 4.8 Division of sample blocks for bird surveys, and the setting of sampling points and line transects

the early morning on sunny weekends based on the daily routines of birds to minimize disturbances by people and transportation to bird activities. Adjusted according to the sunrise time, the survey time was 7:00–10:00 in winter, 6:30–9:30 in spring and autumn, and 6:00–9:00 in summer. During the surveys, we moved forward at a speed of 1.5 km/h, and stayed at each sampling point for 5–20 min based on the scale and characteristics of each habitat. We used 8–10 × binoculars to record the types, numbers, behaviors, and micro-habitats of wild birds seen or heard within 25 m on each side with the line transect as the center. Visible human activities and vehicle disturbances within 25 m of the surrounding area were also recorded (caged birds were recorded in the table, but not included in the calculation of biodiversity variables). Bird behavior includes flying, foraging, calling, resting, etc. Birds were found in such micro-habitats as woodlands, shrubs, ground grasslands, buildings or structures, water bodies, roads, and hard surfaces.

(2) Vegetation surveys

From April to June 2016, we took a comprehensive survey method to conduct vegetation surveys in the study area. We regarded the land blocks divided by roads as the first-level spatial unit in the surveys, and the relevant blocks as second-level survey units by referring to the land boundary signs or the maintenance management unit of green spaces. Scattered green spaces were regarded as separate units. The plants in each survey unit were surveyed and recorded according to type requirements. The background data attributes of the survey elements included the names of green spaces, surveyors and the time of surveys. In the surveys, we recorded the types, names, sizes, numbers, DBH (Diameter at breast height), canopy diameter, crown diameter, area, height, etc. of trees, shrubs, and ground cover plants.

4.1.4 Data Statistics and Analysis Methods

(1) **Calculating the data of bird community and species diversity**

Based on the number of birds and grouped data recorded in the field survey, we calculated the individual number, species richness, diversity index and evenness index of birds in each block.

Individual number refers to the number of birds observed in the study area during the survey period.

Species richness refers to the number of bird species (including subspecies) recorded in the study area, which represents the diversity of wildlife.

Diversity index refers to the numerical value representing the types and quantities of biological communities obtained through the application of mathematical statistics. This study uses the Shannon–Weiner diversity index formula to calculate, namely

$$H' = -\sum (P_i)(\lg P_i)$$

Among them, *Pi* is the ratio of the individual number of the i species to the total number of all species.

Evenness index is used to describe the relative abundance or proportion of individuals in a species. This study uses Pielou's evenness index formula to calculate:

$$J = H' \big/ H_{\max}$$

Among them, H′ has the same meaning as above, $H_{\max} = \ln S$, and S is the number of species.

The superiority index is used to indicate the status and role of the dominant species in a community. In this study, we use the Simpson's diversity index formula to calculate:

$$C = \sum (P_i)^2$$

Among them, Pi has the same meaning as above.

The dominant species is determined according to the encounter rate index *Pi*: dominant species (over 10%), common species (between 1 and 10%), and rare species (less than 1%), indicated by "+++," "++" and "+" respectively.

(2) **Calculating the data of vegetation**

According to the field survey, we calculated the coverage area, coverage ratio, type and height of the trees, shrubs, and ground cover.

The coverage area of trees is calculated according to the following formula:

$$A_t = \sum a_t, \ a_t = \left[(D_i/2)^2 \times \pi \times N\right]$$

where *Di* is the average crown diameter of the tree in the *t* species, and *N* is the number of plants.

The coverage area of shrubs is calculated according to the following formula:

$$A_s = \sum a_s + \sum a_s, \ a_s = \left[(d_s/2)^2 \times \pi \times N\right]$$

where a_s is the area of the bush ball in the *s* species, $a_{s'}$ is the area of the bush ball in the *s* species, $d_{s'}$ is the canopy diameter of bush ball in the *s* species, and *N* is the number of plants. No additional calculation will be made when the projected area of a single bush ball and shrub overlap.

The coverage area of ground cover is calculated according to the following formula:

$$A_h = \sum a_h,$$

where *ah* is the area of ground cover in the *h* species.

For classification and statistics of vegetation types, please refer to the online version of *Flora of China* (http://frps.eflora.cn/). We separately counted the categories of evergreen trees, deciduous trees, all trees, evergreen shrubs, deciduous shrubs, all shrubs, and ground cover, and recorded them by species.

The average height and maximum height of trees and shrubs were counted separately. The average height of trees is calculated according to the following formula:

$$H_t = \sum (a_t \times h_t) \Big/ \sum h_t$$

Among them, h_t is the height of the tree in the *t* species, and a_t has the same meaning as above.

The average height of shrubs is calculated according to the following formula:

$$H_s = \left[\sum (a_s \times h_s) + \sum (a_{s'} \times h_{s'})\right] \Big/ \left(\sum h_s + \sum h_{s'}\right)$$

where h_s is the height of the shrub in the *s* species, $h_{s'}$ is the height of the bush ball in the *s* species, and a_s and $a_{s'}$ have the same meanings as above.

(3) **Calculating the data of urban built environment**

With the help of the AutoCAD map of the study area provided by the Shanghai Pudong New Area Urban Planning and Design Research Institute and the shape file of the current green spaces in Pudong New District, and referring to the revised mapping after the field survey, we first extracted the mapping information of the building, building floor height, green space, waterbody, land parcels and road boundary of each block unit, etc. After importing relevant information into the ArcGIS software, we calculated the variables including land area, floor area, building density, FAR, Green space ratio, water surface ratio, perimeter/area Green space ratios and water bodies, and corrected them based on the high-rise building data of the Shanghai Lujiazui Financial Center provided by Shanghai Lujiazui (Group) Co., Ltd. Afterwards, we converted the green space patch extracted from ArcGIS into Grid format (grid size 1 m × 1 m) and imported it into Fragstats 4.2, a quantitative analysis software designed to compute landscape metrics developed by the Oregon State University, USA, and calculated the landscape indices based on the patch type level as spatial morphology variables.

(4) **Statistical analysis and calculation**

We used statistical analysis software such as Microsoft Excel and IBM SPSS 19.0 FOR WINDOWS to achieve standardized conversion of the bird community structure parameters and the built environment parameters, and explored the influencing mechanism of urban built environment on avian diversity through the Pearson's correlation analysis.

4.2 Bird Communities and Their Structure

4.2.1 Characteristics of Bird Communities

A total of 11,304 bird visits were recorded in the surveys, of which 10,244 were wild birds, 13 were not identified due to insufficient observation time, and the other 10,231 belonged to 47 species in 25 families in 8 orders according to *A Checklist on the Classification and Distribution of the Birds of China* (Zheng 2017). Specifically, there are 20 species of resident birds, 11 species of passing migratory birds, 9 species of winter migratory birds, and 4 species of summer migratory birds, and breeding birds (resident birds and summer visitors) and non-breeding birds (winter visitors and passage migrants) account for approximately 52% and 48% respectively (see Appendix for details). PASSERIFORMES occupy the largest number of birds surveyed, with 38 species in total, and this order alone accounts for 81% of the total, of which there are many small birds such as Muscicapidae (7 species), Turdidae (6 species), Phylloscopidae (3 species), Fringillidae (3 species), and the number of bird species in the other families is less than 2. Among them, there are 6 species birds protected at the municipal level in Shanghai.

Shanghai is located on the northern edge of the oriental realm and adjacent to the palearctic realm. The special geographical location determines that there are few species of resident birds and summer visitors in the bird population structure in Shanghai, while winter visitors and passage migrants are abundant. According to

Fig. 4.9 Bird species in the study area (the top 10 species with the largest numbers are marked with numbers from 1 to 10)

the survey by Luan (2003) of the four types of habitats in Shanghai and Tang et al. (2003) of wild birds in large urban parks and public green spaces in Shanghai's urban area, breeding birds are fewer than non-breeding birds, and most of them are winter visitors and passage migrants. The study area of this research is a high-density urban center area with diversified land-use types and habitat types, and is not part of the main migration routes of migratory birds. The survey results show that the number of breeding birds and non-breeding birds is almost equal, each accounting for about 50%, and that resident birds dominate, which are different from the entire urban area of the city in terms of bird community structure characteristics (Fig. 4.9, Appendix).

Based on the dominance of a single bird species, there are 3 dominant bird species, namely Eurasian Tree Sparrow (*Passer montanus*), Light-vented Bulbul (*Pycnonotus sinensis*) and Chinese Blackbird (*Turdus mandarinus*), accounting for about 2/3 of the total; 7 common species, respectively Spotted Dove (*Streptopelia chinensis*), Black-crowned Night Heron (*Nycticorax nicticorax*), Azure-winged Magpie (*Cyanopica cyanus*), Vinous-throated Parrotbill (*Sinosuthora webbiana*), Mew Gull (*Larus canus*), White Wagtail (*Motacilla alba*), and Long-tailed Shrike (*Lanius schach*), accounting for about 25% of the total. The 10 bird species with the highest dominance, except Mew Gull and White Wagtail, are all resident birds in Shanghai. The other 37 species of birds are rare species with a superiority index less than 1%, accounting for only less than 10% of the total number (Fig. 4.10). As can be seen, the bird community structure of the study area mainly composed of semi-artificial habitats in the attached green spaces is extremely simple.

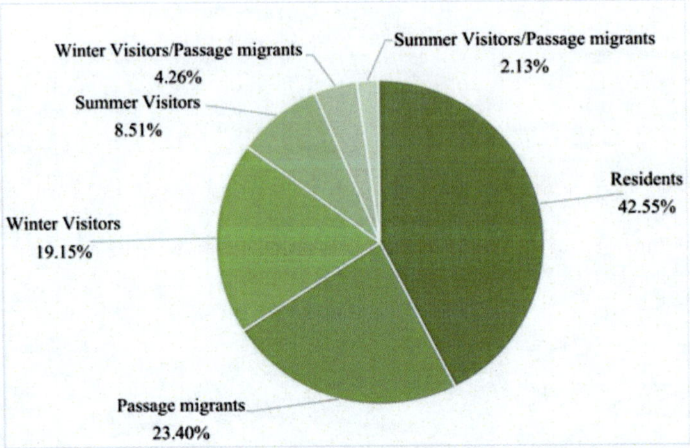

Fig. 4.10 Proportion of resident and migrant types of birds in the study area

4.2.2 *Total Number of Birds, the Number of Bird Species and the Change of Dominant Species in Different Seasons*

Judging from the characteristics of bird communities in the four seasons, there are 38 species in spring, 26 species in summer, 30 species in autumn, and 35 species in winter. The bird community structure shows certain seasonal changes. The total number of birds and the number of bird species recorded in surveys peaked in March and dropped to the lowest in August. And the Pearson correlation coefficient between the number of birds and the number of bird species is as high as 0.945 (p-value = 0.01), showing that the number of birds and the number of bird species basically change synchronously in each season (Fig. 4.11). Eurasian Tree Sparrow, Light-vented Bulbul, and Chinese Blackbird are all dominant species in the four seasons. Black-crowned Night Heron is a dominant species in summer, and Spotted Dove is a dominant species in autumn (Figs. 4.12 and 4.13).

4.2.3 *Comparison of Bird Species Communities in the Study Area and the Century Park in Shanghai*

According to the monitoring reports provided by the Shanghai Wildlife Conservation Management Station, the Shanghai Wild Bird Society of the Shanghai Wildlife Conservation Association and the Century Park Management Office, there are 67 bird species monitored in the Century Park over the years, 38 of which are found in the study area. Nine species of birds including Mew Gull, Brown-eared Bulbul

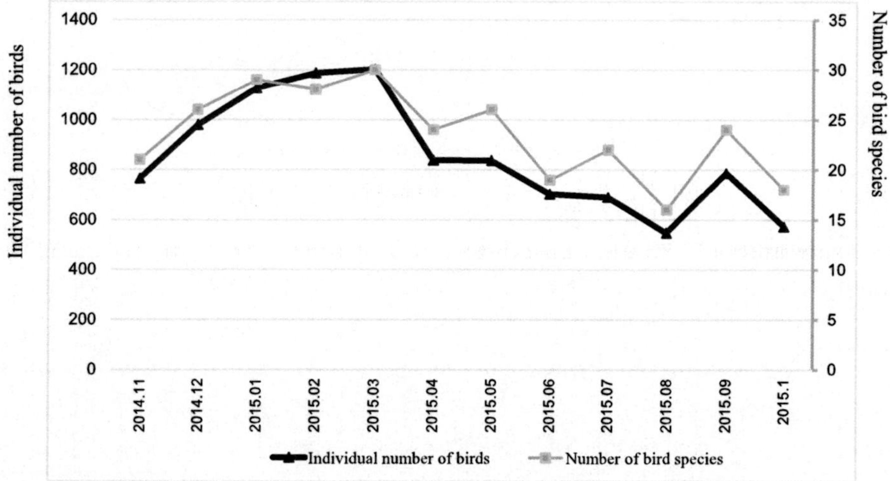

Fig. 4.11 Seasonal changes in the total number of birds and the number of bird species in the study area

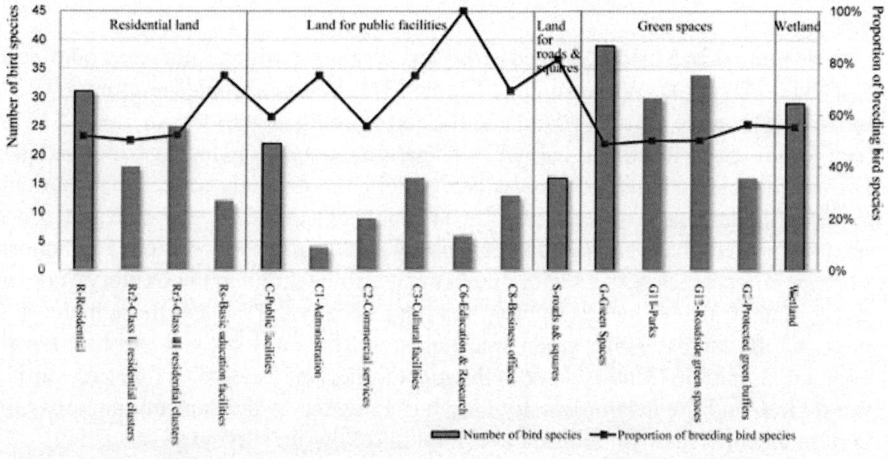

Fig. 4.12 Comparison of the number of bird species and the proportion of breeding bird species of different land-use types in the study area

(*Hypsipetes amaurotis*), Oriental Magpie Robin (*Copsychus saularis*), Narcissus Flycatcher (*Ficedula narcissina*), Chinese Penduline Tit (*Remiz consobrinus*), Plain Prinia (*Prinia inornata*), White-rumped Munia (*Lonchura striata*), Grey-capped Greenfinch (*Chloris sinica*), Yellow-throated Bunting (*Emberiza elegans*) found in the study area have not been recorded in the Century Park monitoring reports before 2016. Among these bird species, except for Mew Gull which only exists by the riverside of Huangpu River, the other 8 are rare species in the study area, some of them

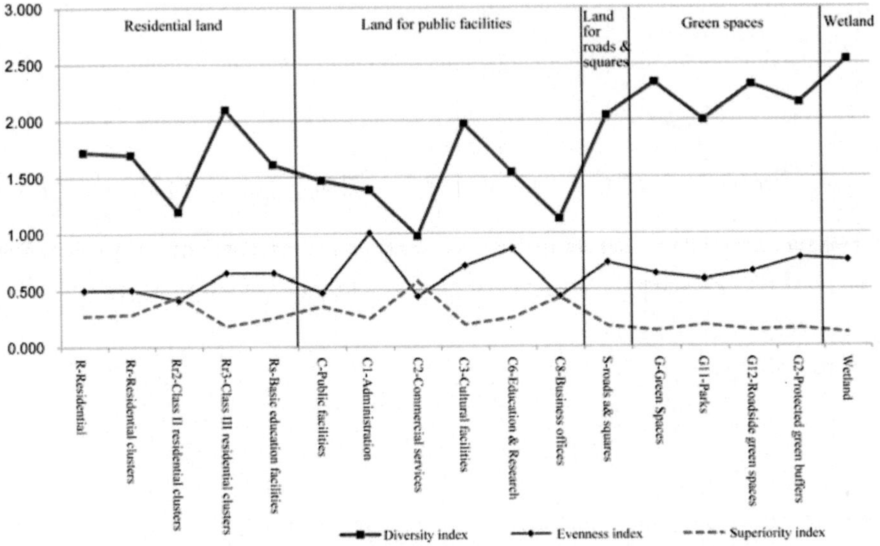

Fig. 4.13 Comparison of the of different land-use types in the study area

with no more than 5 birds recorded in the survey, among which Chinese Penduline Tit (1), Plain Prinia (4), White-rumped Munia (37), Yellow-throated Bunting (10) can only be seen in the small wetland and the surrounding area of the Shanghai Science and Technology Museum as habitats with natural-succession on the abandoned land; Brown-eared Bulbul (4) and Narcissus Flycatcher (1) can only be seen in the Oriental Pearl Park; and Grey-capped Greenfinch (2) only can only be seen in the green space next to the Shanghai Science and Technology Museum subway station. The majority of other bird species with a larger number can also be found in the Century Park, and the dominant species in the Century Park are basically the same as those in the study area. As the largest urban green space close to the study area, it can be basically concluded that the Century Park is the most important "source" habitat for birds in the region, and the avian diversity here has extended to the surrounding areas to a certain extent. This will be further verified in follow-up surveys.

4.2.4 Comparison of the Community Structure of Bird Species in Different Habitats

(1) The number of bird species and the proportion of breeding bird species

By analyzing and comparing the number of bird species, the proportion of breeding bird species, the diversity index, the evenness index, and the superiority index in the different habitat types and different land-use types of the study area, it can be found that the community structure of bird species in various types of land is different.

(a) Among the three major habitats in the study area, the number of bird species of habitats in the urban parks and public green spaces is the largest, with a total of 39 species in 20 families in 6 orders; followed by habitats in the attached green spaces, which have 33 species in 16 families in 3 orders. Although there are only 29 bird species in the wetland habitats, they represent 17 families in 5 orders, and there are more rare bird species. The proportion of breeding bird species (resident birds and summer visitors) in all the three types of habitats exceeds 50%. The diversity index and evenness index of wetland habitats are higher than those of habitats in the urban parks and public green spaces, while the superiority index is the lowest. Habitats in the attached green spaces have the lowest diversity index, but the highest superiority index. The evenness index of all habitats and land-use types are quite low, with the exception of land for roads and squares whose bird community structure is simple.

(b) In habitats in the attached green spaces, the number of bird species of different land-use types is as follows: residential land (31 species in 16 families in 3 orders) > land for public facilities (22 species in 14 families in 3 orders) > land for roads and squares (16 species in 13 families in 3 orders). The proportion of breeding bird species (resident birds and summer migratory birds) in the three land-use types is higher than 50%: land for roads and squares (81.25%) > land for public facilities (59.09%) > residential land (51.61%). In terms of the diversity index: land for roads and squares (S) > residential land (R) > land for public facilities (C); In terms of the evenness index: land for roads and squares (S) > residential land (R) > land for public facilities (C); In terms of the superiority index: land for public facilities (C) > residential land (R) > land for roads and squares (S). The above data indicates that the greater the degree of human disturbances is, the simpler the community structure of bird species will be.

(c) Regarding residential land, although the FAR ofland for Class II residential clusters (R2), which is mainly multi-storey housing, is lower than that of land for Class III residential clusters (R3), which is mainly high-rise housing, the number of bird species in the attached residential green spaces of land for Class II residential clusters (R2) is simpler, of which the diversity index and evenness index are lower, and the superiority index is higher; The number of bird species in land for basic education[1] facilities (Rs) is less than that of the above two land-use types, while its proportion of breeding bird species is much higher than that of lands for Class II and Class III residential clusters. The diversity index and the superiority index here between those of lands for Class II and Class III residential clusters, and the evenness index is basically the same as that of land for Class III residential clusters.

(d) The number of bird species of land for public facilities, from most to least: land for cultural facilities (C3) > land for business offices (C8) > land for commercial

[1] Basic education in China equals K-12 education.

services (C2) > land for education and research (C6) > land for administration (C1); the proportion of breeding birds of land for public facilities is generally higher than that of other land-use types. In fact, the birds found in land for education and research are all resident birds. In terms of the diversity index: land for cultural facilities (C3) > land for education and research (C6) > land for administration (C1) > land for business offices (C8) > land for commercial services (C2). In terms of the evenness index, land for administration (C1) and land for education and research (C6) are the highest, followed by land for cultural facilities (C3). Land for business offices (C8) and land for commercial services (C2) have the lowest diversity index and evenness index, but the highest superiority index. This is because land for cultural facilities, land for education and research, and land for administration generally have a certain scale of green spaces inside or at the entrance and exit, while the attached green spaces of land for business offices and land for commercial services are smaller in scale and subject to more artificial disturbances. Therefore, land for cultural facilities, land for education and research, and land for administration are more attractive for birds to rest and forage, indicating that the more artificial an environment is, the larger its proportion of bird dominant species will be, that is, the simpler its community structure of bird species will be.

(e) There is no significant difference in the proportion of breeding bird species of various green spaces, basically around 50%, while the number of bird species of protected green buffers is significantly less than that of the former two. Roadside green spaces, especially riverside green spaces, thanks to their location at the intersection of two types of habitats, have a larger number of bird species, a higher diversity index and a higher evenness index than those of parks, and its diversity index is also higher than that of protected green buffers. The superiority index of all types of green spaces is basically the same.

(2) **Dominant species and common species**

When it comes to the dominance of a single bird species, the dominant species of wetland habitats is slightly different from those of the other two types of habitats (semi-natural habitats in the urban parks and public green spaces, and semi-artificial habitats in the attached green spaces). In urban parks and public green spaces, and in residential land and land for public facilities in semi-artificial habitats in the attached green spaces, except for Eurasian Tree Sparrow and Light-vented Bulbul, which are both dominant species, other dominant species and common species are different in various land-use types, but in all types of semi-artificial habitats in the attached green spaces, the dominant species are resident birds, including Chinese Blackbird, Spotted Dove, Black-crowned Night Heron, Vinous-throated Parrotbill, etc., and the common species are also mainly resident birds, such as Azure-winged Magpie, Silky Starling (*Spodiopsar sericeus*), Cinereous Tit (*Parus cinereus*), Long-tailed Shrike, etc. The dominant species and common species of protected green buffers are also different from that of other land-use types. Apart from the two resident birds, Chinese Blackbird and Azure-winged Magpie, the dominant species also include winter visitors such as Dusky Thrush (*Turdus eunomus*). Winter migratory birds and passage migrants such as Eyebrowed Thrush (*Turdus obscurus*), Olive-backed Pipit (*Anthus hodgsoni*), White's Thrush (*Zoothera aurea*), and Pale Thrush (*Turdus pallidus*) also belong to common species.

Therefore, the bird species attracted by wetland habitats and protected green buffers are different from those attracted by other types of habitats, which is of special value to the avian diversity in urban centers. See Tables 4.1 and 4.2 for details.

4.2.5 Comparison of Bird Community Structure in Various Morphologies of Semi-artificial Habitats in the Attached Green Spaces

The changes in urban green space morphology have remarkable influence on bird communities. The green space-building morphology determines to a certain extent the scale, connectivity, and the degree of interference caused by frequent human use. Therefore, this study further analyzes the differences in bird species communities in various green space-building morphologies in different land-use types, and provides a basis for the rational layout of green spaces. Take various land-use types of semi-artificial habitats in the attached green spaces for example, the habitats of sampling points are divided into the following four morphologies based on the relationship between the green space habitats where the sampling points are located in and the buildings (Fig. 4.14).

(1) Multi-storey residential areas, which is defined as land for Class II residential clusters (Rr2) in the classification of land use types: central green spaces (Rr2C), green spaces between houses (Rr2H), roadside green spaces (Rr2S);
(2) High-rise residential areas, which is defined as land for Class III residential clusters (Rr3): central green spaces (Rr3C), green spaces between houses (Rr3H), roadside green spaces (Rr3S);
(3) Business, which is defined as land for business offices (C8): central green spaces (C8C), peripheral green spaces (C8B), entrance green spaces (C8E);
(4) Commercial areas, which is defined as land for commercial services (C2): central green spaces (C2C), peripheral green spaces (C2P).

The sampling points are classified according to the 11 types of green spaces mentioned above. The analysis results (Figs. 4.15 and 4.16) show that in residential areas, the number and diversity index of bird species in central green spaces (Rr3C) in high-rise residential areas are much larger than those in other types, while the proportion of breeding bird species is the smallest, the dominance is the lowest, and the evenness index is the second high, with the richest bird community structure; Although green spaces between houses (Rr3H) in high-rise residential areas have the highest evenness index, the number of bird species is the smallest and the proportion

Table 4.1 Dominant and common bird species of different land-use types in the study area

	Residential land (R)	Including			Land for public facilities (C)	Including			
		Land for Class II residential clusters (Rr2)	Land for Class III residential clusters (Rr3)	Land for basic education facilities (Rs)		Land for commercial services (C2)	Land for cultural facilities(C3)	Land for education and research (C6)	Land for business offices (C8)
Number of bird species	31	18	25	12	22	9	16	6	13
Dominant species	Eurasian Tree Sparrow, Light-vented Bulbul, Chinese Blackbird	Eurasian Tree Sparrow, Light-vented Bulbul	Light-vented Bulbul, Eurasian Tree Sparrow, Chinese Blackbird, Spotted Dove	Eurasian Tree Sparrow, Light-vented Bulbul, Chinese Blackbird	Eurasian Tree Sparrow, Light-vented Bulbul	Eurasian Tree Sparrow, Light-vented Bulbul	Eurasian Tree Sparrow, Light-vented Bulbul, Chinese Blackbird	Light-vented Bulbul, Azure-winged Magpie, Eurasian Tree Sparrow, Spotted Dove	Eurasian Tree Sparrow, Light-vented Bulbul
Common species	Spotted Dove, Azure-winged Magpie, Vinous-throated Parrotbill, Silky Starling, White Wagtail	Chinese Blackbird, Spotted Dove, Azure-winged Magpie	Vinous-throated Parrotbill, Azure-winged Magpie, Silky Starling, White Wagtail, Yellow-browed Warbler (*Phylloscopus inornatus*), Cinereous Tit	Spotted Dove, White Wagtail, Long-tailed Shrike, Pale Thrush, Cinereous Tit	Chinese Blackbird, Spotted Dove, White Wagtail, Azure-winged Magpie	Spotted Dove, Yellow-browed Warbler, Chinese Blackbird, White Wagtail	Spotted Dove, White Wagtail, Black-crowned Night Heron, Barn Swallow (*Hirundo rustica*), Dusky Thrush, Vinous-throated Parrotbill, Azure-winged Magpie	Chinese Blackbird, Long-tailed Shrike	Chinese Blackbird, Spotted Dove

(continued)

Table 4.1 (continued)

	Land for roads and squares (S)	Green spaces (G)	Of which			Wetlands (land reserves for public facilities Xc)
			Parks (G11)	Roadside green spaces (G12)	Protected green buffers (G2)	
Bird species	16	39	30	34	16	29
Dominant species	Light-vented Bulbul, Eurasian Tree Sparrow, Chinese Blackbird, Spotted Dove	Eurasian Tree Sparrow, Light-vented Bulbul, Chinese Blackbird, Black-crowned Night Heron	Black-crowned Night Heron, Eurasian Tree Sparrow, Chinese Blackbird, Light-vented Bulbul	Light-vented Bulbul, Eurasian Tree Sparrow, Chinese Blackbird	Chinese Blackbird, Azure-winged Magpie, Dusky Thrush	Eurasian Tree Sparrow, Light-vented Bulbul, Vinous-throated Parrotbill
Common species	Vinous-throated Parrotbill, Azure-winged Magpie, Long-tailed Shrike, Crested Myna (*Acridotheres cristatellus*), White Wagtail, Cinereous Tit, Barn Swallow, Daurian Redstart (*Phoenicurus auroreus*)	Spotted Dove, Mew Gull, Azure-winged Magpie, Vinous-throated Parrotbill, Little Egret, Dusky Thrush, Long-tailed Shrike, White Wagtail	Spotted Dove, Vinous-throated Parrotbill, Long-tailed Shrike	Spotted Dove, Black-crowned Night Heron, Azure-winged Magpie, Little Egret, Long-tailed Shrike, Yellow-browed Warbler, White Wagtail, Pale Thrush	Spotted Dove, Eurasian Tree Sparrow, Light-vented Bulbul, Eyebrowed Thrush, Olive-backed Pipit, White's Thrush, Long-tailed Shrike, Pale Thrush	Chinese Blackbird, Common Moorhen, Azure-winged Magpie, White-rumped Munia, Spotted Dove, Dusky Thrush, Crested Myna, Little Grebe (*Tachybaptus ruficollis*), Long-tailed Shrike, Yellow-throated Bunting, Black-crowned Night Heron

Table 4.2 Dominant species in different green space-building morphologies in different land-use types of the study area

	Central green spaces	Green spaces between houses	Roadside green spaces	Peripheral green spaces	Entrance green spaces
Rr2	Eurasian Tree Sparrow, Light-vented Bulbul	Eurasian Tree Sparrow, Light-vented Bulbul	Eurasian Tree Sparrow, Light-vented Bulbul, Chinese Blackbird	–	–
Rr3	Eurasian Tree Sparrow, Light-vented Bulbul, Chinese Blackbird, Spotted Dove	Eurasian Tree Sparrow, Light-vented Bulbul, Grey-backed Thrush (*Turdus hortulorum*)	Light-vented Bulbul, Eurasian Tree Sparrow, Chinese Blackbird	–	–
C8	Eurasian Tree Sparrow, Light-vented Bulbul	–	–	Eurasian Tree Sparrow, Light-vented Bulbul	Eurasian Tree Sparrow, Light-vented Bulbul
C2	Eurasian Tree Sparrow, Light-vented Bulbul	–	–	Eurasian Tree Sparrow, Light-vented Bulbul	–

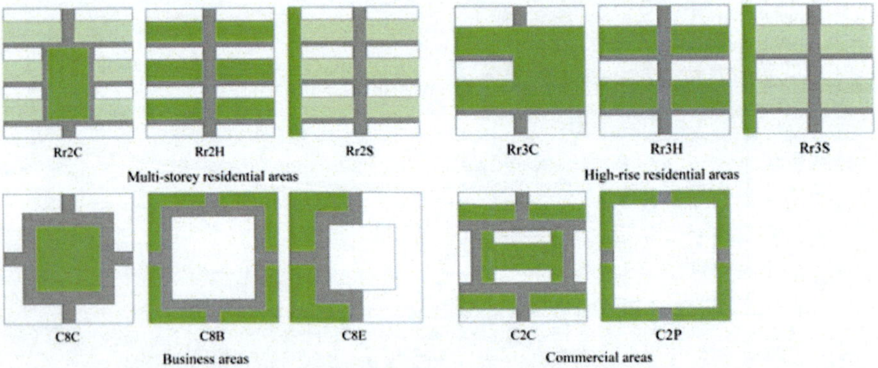

Fig. 4.14 Illustration of green space-building morphology in various land-use types of semi-artificial habitats in the attached green spaces

of breeding bird species is the highest. The number of bird species in green spaces between houses (Rr2H) in the multi-storey residential area ranks second among all types of green spaces in residential land, but the diversity index and evenness index are the lowest, and the superiority index is the highest, which indicates that the bird species here are mostly several dominant species such as Eurasian Tree Sparrow and Light-vented Bulbul. It can be seen that high-intensity development does not

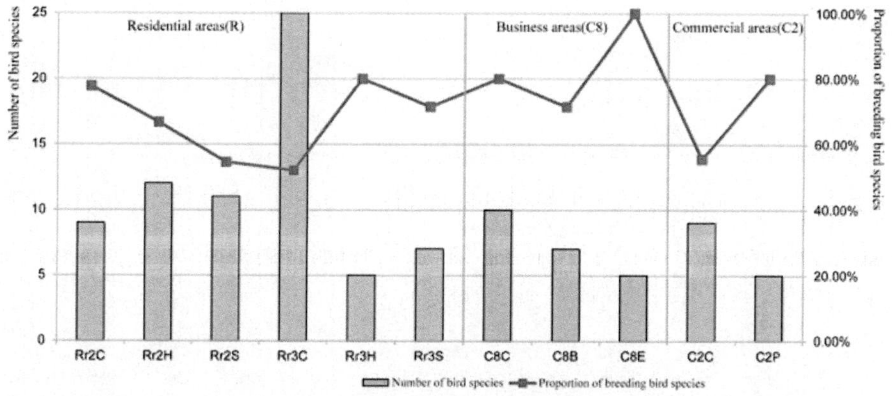

Fig. 4.15 Comparison of the number of bird species and the proportion of breeding bird species in various green space-building morphologies in different land-use types

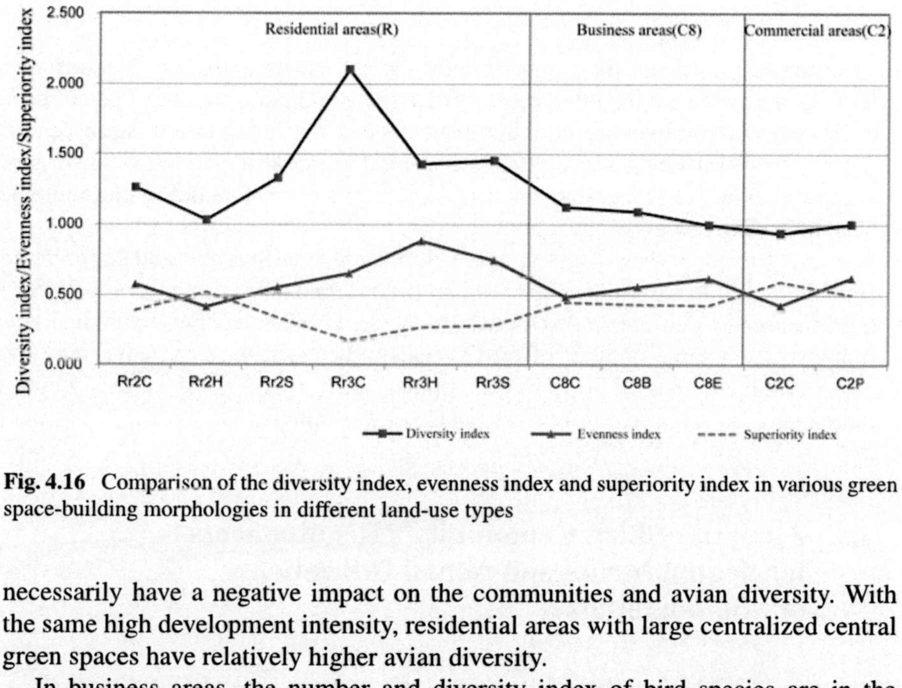

Fig. 4.16 Comparison of the diversity index, evenness index and superiority index in various green space-building morphologies in different land-use types

necessarily have a negative impact on the communities and avian diversity. With the same high development intensity, residential areas with large centralized central green spaces have relatively higher avian diversity.

In business areas, the number and diversity index of bird species are in the following descending order: central green spaces (C8C) > surrounding green spaces (C8B) > entrance green spaces (C8E). The proportion of breeding bird species is highest in peripheral green spaces (100%); their evenness index and superiority index in various green spaces are not much different, while the superiority index is similar to the relevant data in central green spaces in multi-storey residential areas. Central green spaces also show certain advantages in supporting avian diversity.

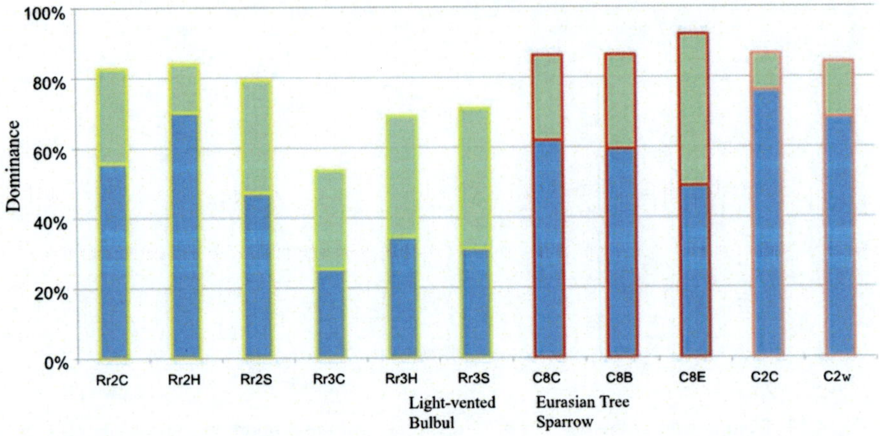

Fig. 4.17 Changes in the dominance of Eurasian Tree Sparrow and Light-vented Bulbul in different habitat types in the study area

In commercial areas, the number of bird species and their superiority index are in the following descending order: central green spaces (C2C) > peripheral green spaces (C2P), whereas the diversity index and evenness index are in the following descending order: peripheral green spaces (C2P) > central green spaces (C2C). Since green spaces in commercial areas are small, different green space-building morphologies have little effect on avian diversity.

Among the 11 types of green space, Eurasian Tree Sparrow and Light-vented Bulbul are absolute dominant species, but their dominance is significantly lower in high-rise residential areas, that is to say, in blocks with a better ecological environment and a large number of bird species, other birds have an advantage over urban dominant species birds in the competition for ecological niches, which is more beneficial to the overall diversity (Fig. 4.17, Table 4.2).

4.3 Analysis of Bird Community's Requirements for Spatial Niches and Actual Utilization of Micro-habitats

4.3.1 Analysis of Bird Community's Requirements for Spatial Niches

The habitat selection of birds depends on whether a habitat can provide them with sufficient food resources, suitable breeding sites, conditions for avoiding natural enemies and bad weather, etc., to ensure the survival and reproduction of bird species (Jiang et al. 2012). Behind these preferences for habitat selection are the requirements for niches. Food resources and breeding sites are the most important niches for urban wild birds. We can determine the requirements of birds in the study area for feeding and nesting niches based on their feeding and nesting habits.

(1) **Requirements for feeding niches**

Feeding is the first and foremost requirement for survival of wildlife. The urban habitats, which are quite complex, also provides abundant food resources for birds. With reference to the research by Wang et al. (2000), we divide the birds in the study area into the following 5 types according to their feeding groups.

(1) Insectivorous: mainly feed on insects and their larvae;
(2) Omnivorous: feed on both insects and plants;
(3) Herbivorous: feed on plants;
(4) Carnivorous: feed on vertebrates such as birds, frogs, and mammals;
(5) Piscivorous: feed on intertidal organisms such as fish and shrimp.

Among the 47 species of birds recorded, in terms of the number of bird species, the insectivorous bird species make up the largest proportion (36%), followed by the omnivorous bird species (32%); calculated by the number of birds, the omnivorous birds account for as high as 84%, which is consistent with the higher proportion of omnivorous animals in the "synurbization" evolution of urban wildlife (Fig. 4.18).

Through further analysis, we can see that the bird species recorded in the survey mainly feed on the ground, bushes, and shrubs. In terms of quantity, the omnivorous birds that feed on the ecological space of the bushes/ground make up the largest proportion, followed by ground-feeding birds. Therefore, the requirements for feeding niches of bird species in the study area are mainly bushes and ground. In addition, piscivorous water birds such as the wading birds including Black-crowned Night Heron (*Nycticorax nicticorax*), Little Egret (*Egretta garzetta*) and Chinese Pond Heron (*Ardeola bacchus*), and the swimming birds including Little Grebe (*Tachybaptus ruficollis*), Common Moorhen (*Gallinula chloropus*) and Mew Gull (*Larus canus*), only need water bodies (including intertidal zones) as their feeding niches (Fig. 4.19).

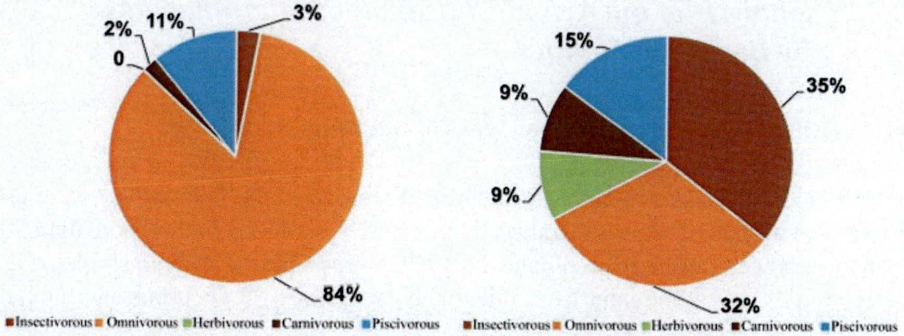

Fig. 4.18 Proportions of feeding habits of bird communities in the study area (left: calculated by the number of bird species; right: calculated by the number of birds)

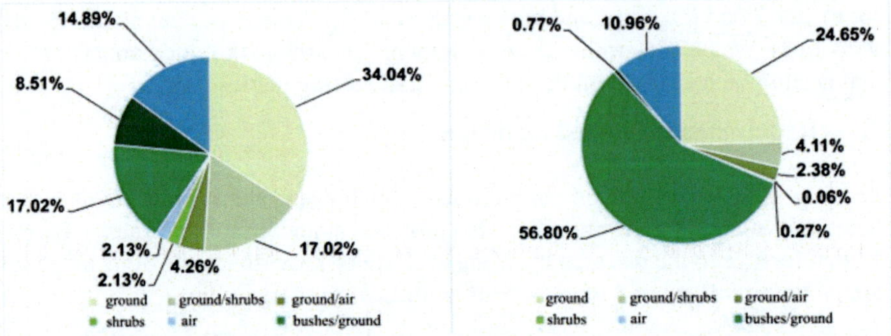

Fig. 4.19 Requirements of birds for feeding niches in the study area (left: calculated by the number of bird species; right: calculated by the number of birds)

(2) Requirements for nesting niches

Among the 47 species of birds recorded in the survey, there are 25 species of breeding birds. According to the different nesting environments of breeding birds and referring to the research of Chen et al. (2000), the nesting niches of birds can be divided into floating nests, ground nests, shrub nests, tree nests, tree hole or cavity nests, building nests, etc.

Research on the breeding niches of birds shows that the breeding bird species in the study area prefer tree nests and building nests, with the proportions being 42% and 18% respectively; 16% of the breeding bird species prefer shrub nests and tree hole or cavity nests; and only 8% of them prefer floating nests. By studying the 25 breeding bird species, we find that that tree nests and building nests account for the largest proportions (both 47%), and other types of nests account for less than 5% respectively. Floating nests and tree hole or cavity nests are the least (1%) (Fig. 4.20). Therefore, the nesting niches of bird species in the study area are highly dependent on forest resources and artificial environment.

4.3.2 Analysis of the Actual Utilization of Micro-habitats by Bird Communities

(1) Utilization ratios of various types of micro-habitats

In this study, the micro-habitats of the birds recorded in the field survey were used to analyze the actual space utilization by them. The results show that more than 50% of the birds were found in woodlands, 8.47% in grasslands, 4.22% in shrubs, 5% in water bodies, and at the same time, due to the particularity of the urban environment,

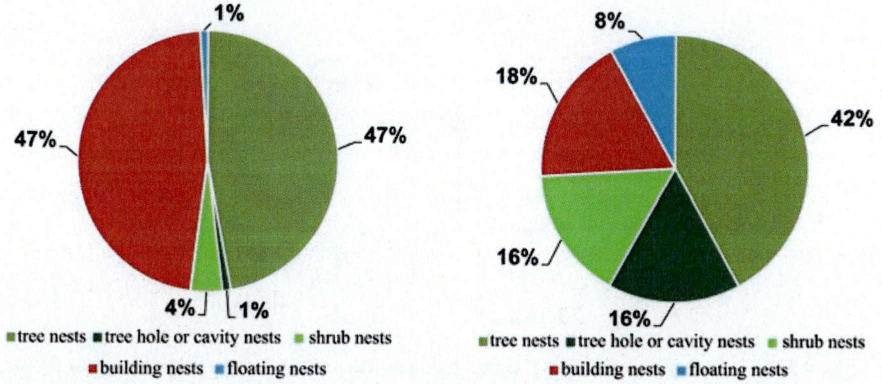

Fig. 4.20 Proportions of nesting niches of breeding birds in the study area (left: calculated by the number of bird species; right: calculated by the number of birds)

10.85% on the surface of buildings or structures, 4.41% on sealed pavements, and 2.23% on roads. Trees, ground, and buildings are the most utilized micro-habitats (Fig. 4.21).

(2) Comparison with requirements for spatial niches

Compared with the requirements for spatial niches in Sect. 4.3.1, the results show that the foraging behaviors of bird communities are mainly in woodlands and grasslands, and occur in areas such as firm ground, waters, and roads. A total of 11 foraging behaviors of birds were recorded, involving 36 birds in 8 species, including White Wagtail, Light-vented Bulbul, Daurian Redstart (*Phoenicurus auroreus*), Chinese Grosbeak (*Eophona migratoria*), Yellow-browed Warbler (*Phylloscopus inornatus*), Azure-winged Magpie, Silky Starling and Long-tailed Shrike. The food they forage includes camphor fruit, crabapple fruit, Ligustrum lucidum fruit, and other fruits, as

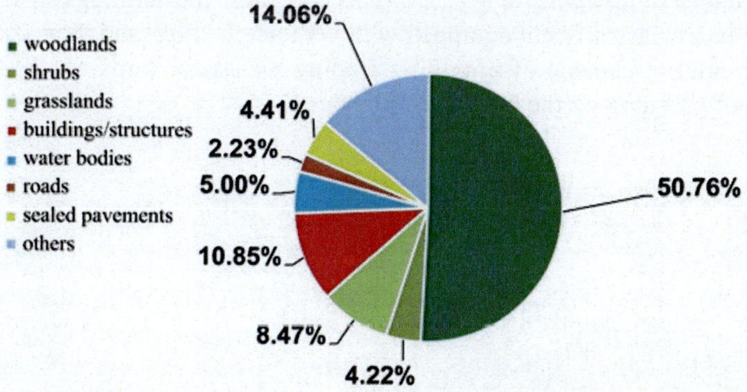

Fig. 4.21 The actual utilization ratios of various micro-habitats by bird communities in the study area

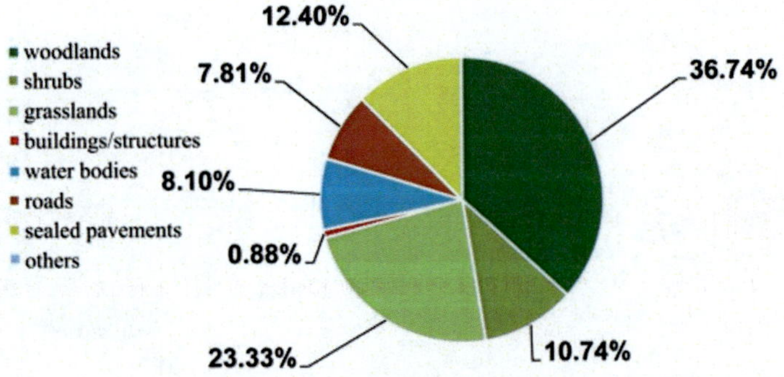

Fig. 4.22 The actual distribution of micro-habitats where bird communities forage in the study area

well as invertebrates such as earthworms and small birds, and their feeding spaces are in the woodlands or on the ground. This is basically consistent with the requirements of bird species for feeding niches (Fig. 4.22).

The nesting behavior of birds was hardly seen during the survey. Only hundreds of Black-crowned Night Heron nests were found in the Oriental Pearl Park, and some Light-vented Bulbul nests and Eurasian Tree Sparrow nests were found in individual residential areas (Fig. 4.23). Black-crowned Night Heron nests and Light-vented Bulbul nests are branch nests over 3 m above the ground, while Eurasian Tree Sparrow nests are nests in the holes in buildings. This partly justifies the hypothesis about their requirements for nesting niches.

(3) Special urban micro-habitats–buildings/structures

Except for buildings/structures, the proportions of birds behaviors in other types of micro-habitats are similar, and the proportion of foraging behaviors of birds is only slightly higher on grassland and firm ground. Buildings/structures mainly provide resting places for birds, and the proportion of birds' foraging behaviors occurring there declines significantly (Fig. 4.24). This is because the buildings/structures in the study area are basically not equipped with vertical planting, and their vertical space surfaces are not capable of providing feeding niches for birds except for a small amount of greening on the balconies of residents.

Fig. 4.23 A Black-crowned Night Heron nest in the Oriental Pearl Park, a Light-vented Bulbul nest in Meiyuan New Village, and a Eurasian Tree Sparrow nest in Laoshan New Village

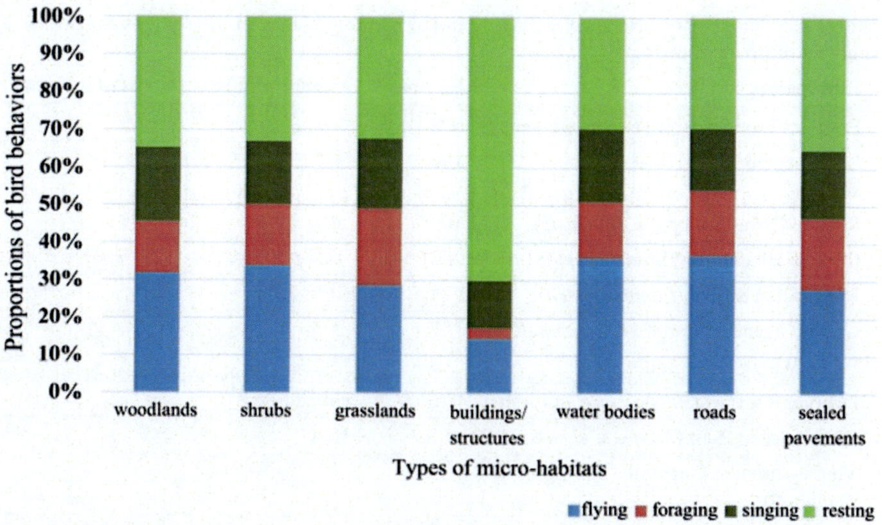

Fig. 4.24 Proportions of bird behaviors in various micro-habitat spaces in the study area

A total of 994 birds of 15 species were found on buildings or structures, and the numbers of floors on which 435 birds were found was recorded. We arrange the bird species involved by number in descending order: Eurasian Tree Sparrow, Light-vented Bulbul, Spotted Dove, Chinese Blackbird, etc. That is to say, a large number of dominant species were found to stay on buildings or structures. Among the seven common species, except for Black-crowned Night Heron and Mew Gull, which rely on water bodies for survival, the other five common species were found to stay on buildings or structures (Table 4.3).

The analysis of the number and species of birds staying in buildings/structures micro-habitats in various land-use types shows that the number of birds staying in the buildings in land for Class II residential clusters (Rr2) is the largest. In this type of residential area with older buildings and a higher proportion of elderly residents, the proportion of building balconies and platforms that remain open and are planted with greenery is usually higher than in that in land for Class III residential clusters (Rr3), and therefore more attractive to birds for feeding and habitation. The number of birds staying in the annexes in public green spaces ranks second. In terms of the number of bird species, public green spaces rank first, followed by land for Class II residential clusters, land for basic education facilities, and then Land for Class III residential clusters (Fig. 4.25).

With respect to the vertical spatial niches of the buildings where they stay, the lowest floor recorded by us for birds in the study area to stay on is the half floor, the highest the 16th floor, and the average number four and a half floors. More than 90% of the birds stay below the 6th floor, and most of them stay on the 5th floor. The number of birds staying on the 7th–16th floors was basically the same, but there was a small peak in the number of birds staying on the 12th floor (Fig. 4.26).

Table 4.3 Number of bird species staying in the building niches in the study area and average number of floors of their stay

Bird species	Number	Average number of floors of their stay
Eurasian Tree Sparrow (*Passer montanus*)	608	4.1
Light-vented Bulbul (*Pycnonotus sinensis*)	186	5.7
Spotted Dove (*Spilopelia chinensis*)	100	4.3
Chinese Blackbird (*Turdus merula*)	53	4.8
Long-tailed Shrike (*Lanius schach*)	8	4.5
Crested Myna (*Acridotheres cristatellus*)	7	5.5
White Wagtail (*Motacilla alba*)	6	5
Daurian Redstart (*Phoenicurus auroreus*)	6	Unrecorded
Olive-backed Pipit (*Anthus hodgsoni*)	5	10.4
Vinous-throated Parrotbill (*Sinosuthora webbiana*)	5	2
Silky Starling (*Spodiopsar sericeus*)	4	2.7
Cinereous Tit (*Parus major*)	3	4.5
Azure-winged Magpie (*Cyanopica cyanus*)	1	Unrecorded
Oriental Magpie Robin (*Copsychus saularis*)	1	3
Common Magpie (*Pica pica*)	1	Unrecorded
Total	994	4.5

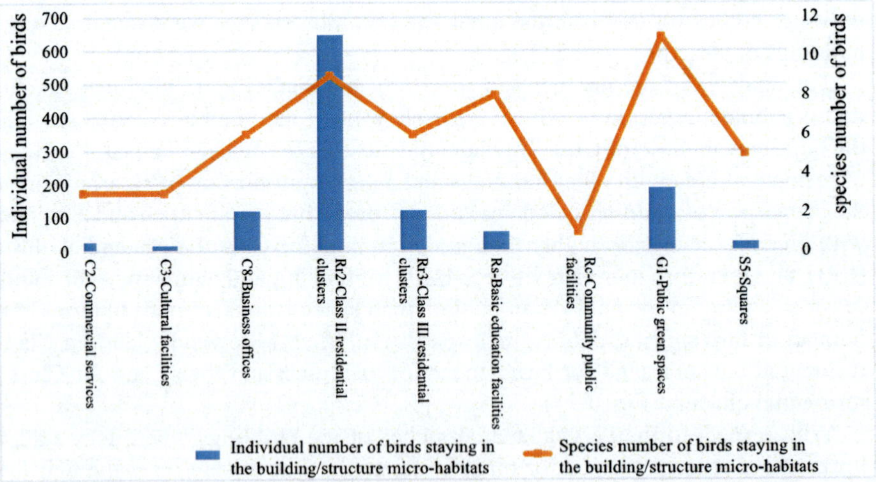

Fig. 4.25 Individual number and species number of birds staying in the building/structure micro-habitats of various land-use types in the study area

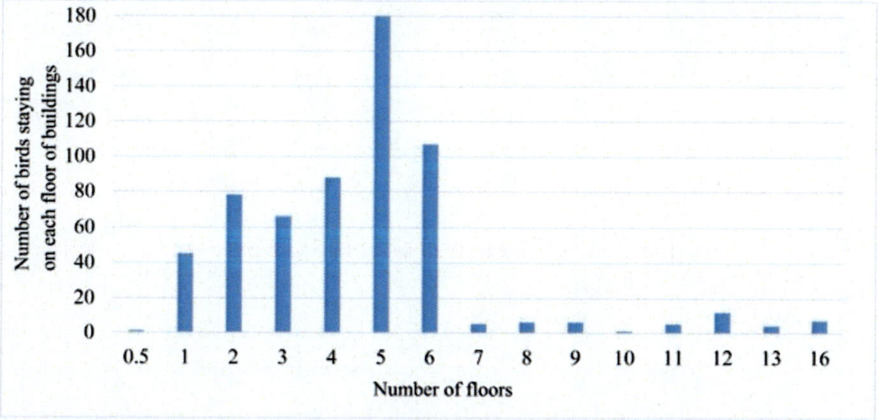

Fig. 4.26 Number of birds staying in the micro-habitats on each floor of buildings

Through further analysis of the numbers of floors where birds stay in various types of micro-habitats in various land-use types, we find that: in public green spaces and the land for squares, most birds stay on the 1st–2nd floors of buildings. This is related to the numbers of floors of the annex buildings in the public green spaces and squares. For land for Class II residential clusters, land for commercial services, land for cultural facilities, and infrastructure land where the numbers of floors of buildings generally does not exceed 6, the floors where birds stay on commercial buildings and cultural buildings are lower (below the 3rd floor), while in primary and middle school buildings and buildings in land for Class II residential clusters, they generally stay on about the 5th floor. For land for business offices and Land for Class III residential clusters where the numbers of floors of buildings exceeds 6, the floors where birds stay on commercial buildings are also not above the 6th floor. This may be because the walls of high-rise commercial buildings in the study area are mainly glass curtain walls, lacking stepping stones for birds to climb up. In the small high-rise and high-rise residential buildings in Land for Class III residential clusters, the proportion of birds staying on each floor is relatively balanced, mostly below the 3rd floor and on the 12th floor (Fig. 4.27), indicating the possibility of birds using vertical spaces as their off-ground habitats.

The analysis of the vertical spatial distribution characteristics of birds' staying behaviors on buildings shows that flying and foraging behaviors generally occur below the 4th floor, and resting behaviors may occur in higher vertical spaces, but mainly below the 7th floor (Fig. 4.28).

Apart from the birds staying on the buildings, another 56 birds were recorded as staying on electric wires, telephone poles, light poles, street signs, construction site scaffolding, and billboards. Most of them were recorded as staying on telephone poles and electric wires (Fig. 4.29).

The study shows that some birds rest and forage on the surface of buildings/structures, while some other birds that take tree-holes as their habitats in the

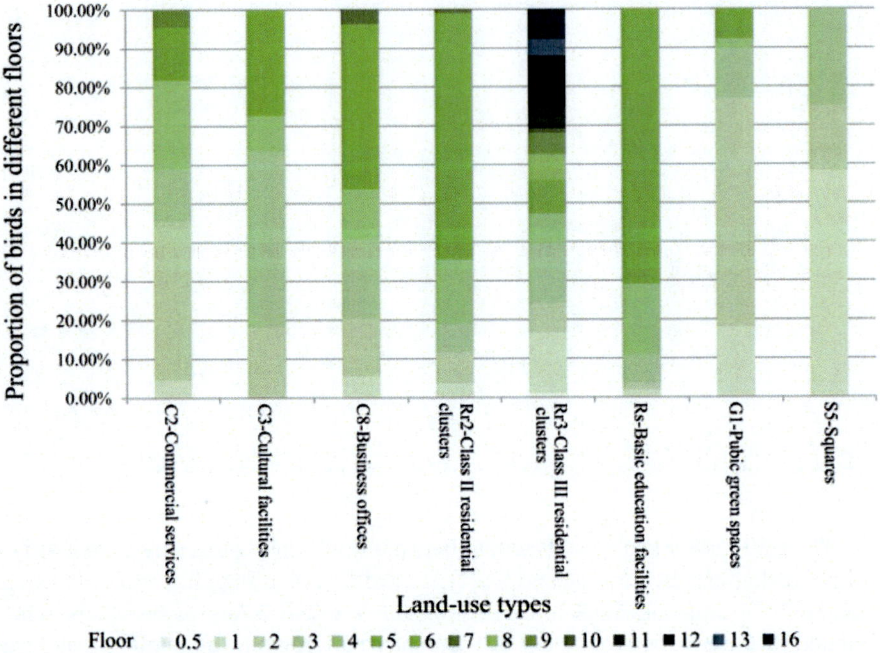

Fig. 4.27 Vertical spatial distribution of the number of birds staying in the micro-habitats of buildings in various land-use types in the study area

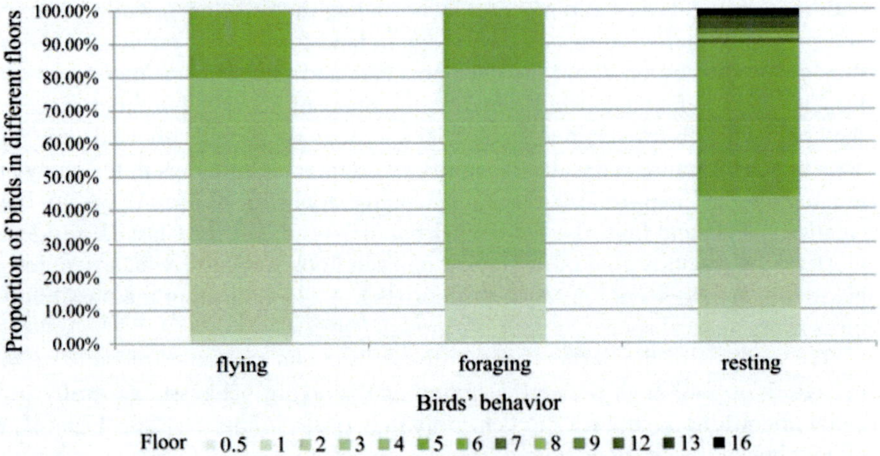

Fig. 4.28 Vertical spatial distribution of birds staying on the buildings in the study area

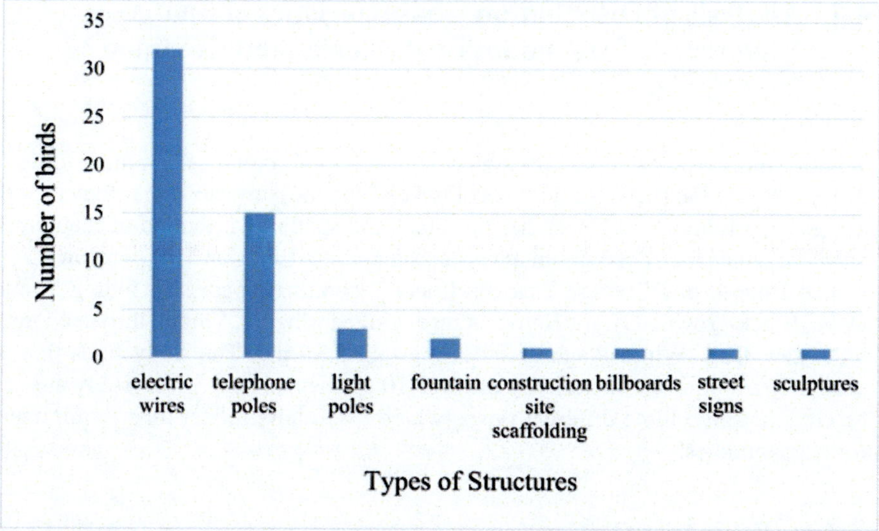

Fig. 4.29 Number of birds staying on various structures in the study area

pristine environment have a certain space selection preference for holes in buildings and urban sculptures, street lights, billboards, and even cross arms of traffic lights at intersections in the urban environment. These holes can shelter them from wind and rain. Whether they are nesting here for a long time or resting for a short time, they feel safe and sheltered, and that these places are suitable for nesting (Fig. 4.30). Of course, this behavior of birds inevitably causes certain negative impacts on the human settlements. For example, there are several gas poisoning accidents and even deaths caused by sparrows nesting in the pipes of water heaters every year in Shanghai. But from another perspective, in order to avoid the tragedies caused by birds "occupying" the ventilation ducts of the human settlements as habitats, it is better for humans to take the initiative to provide suitable spaces for birds to build nests.

Fig. 4.30 Birds' space selection preference for holes in the buildings/structures of the study area

4.4 Analysis of Distribution Characteristics of Spatial Preference of Dominant Bird Species and Common Bird Species

Wildlife has certain demand preferences in terms of spatial selection. In order to further discuss the spatial preference of the birds in the study area, the study uses the Power Map function of Excel 2013 to block the spatial distribution of the number and dominance index of 3 dominant species including Eurasian Tree Sparrow, Light-vented Bulbul, and Chinese Blackbird, and 7 common species including Spotted Dove, Black-crowned Night Heron, Azure-winged Magpie, Vinous-throated Parrotbill, Mew Gull, White Wagtail, and Long-tailed Shrike. The study finds that the spatial distribution characteristics of these 10 species of birds in the study area can be classified into three different types, which essentially reflect their requirements for spatial niches.

4.4.1 Even Distribution

Overall, the spatial selection of dominant species including Eurasian Tree Sparrow, Light-vented Bulbul, and Chinese Blackbird, and Spotted Dove with the largest number in common species features even distribution. These four birds are the most abundant bird species in Shanghai. They are mainly omnivorous and arboreal, and their nesting environment includes trees/buildings. Since they are highly adaptable to artificial environment, these birds are large in number, have a high degree of dominance, and are evenly distributed in various blocks with different land-use types, different spatial structures, and different density gradients. But in the small wetland of the Shanghai Science and Technology Museum with the highest bird species richness, the number and dominance of Eurasian Tree Sparrow, a bird species with the highest dominance overall, are lower than in other blocks (Figs. 4.31, 4.32, 4.33 and 4.34). Therefore, in habitats with abundant biological species, species that are highly adapted to urban habitats do not necessarily have an advantage in niche competition.

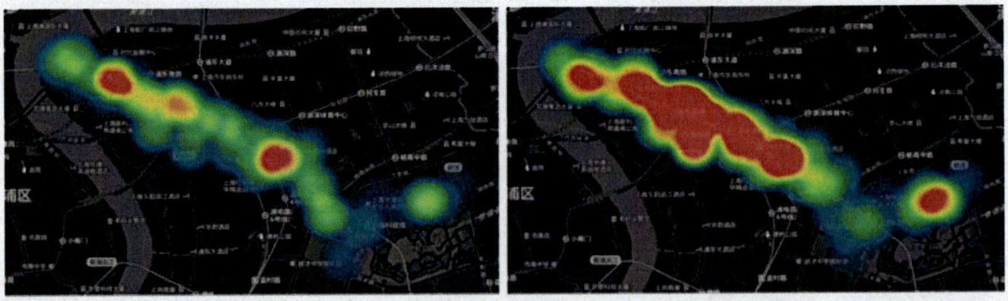

Fig. 4.31 Distribution of the number and dominance of Eurasian Tree Sparrow in the study area (left: number; right: dominance)

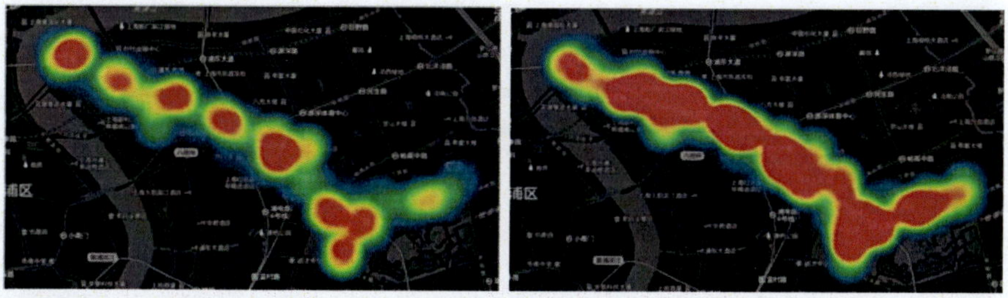

Fig. 4.32 Distribution of the number and dominance of Light-vented Bulbul in the study area (left: number; right: dominance)

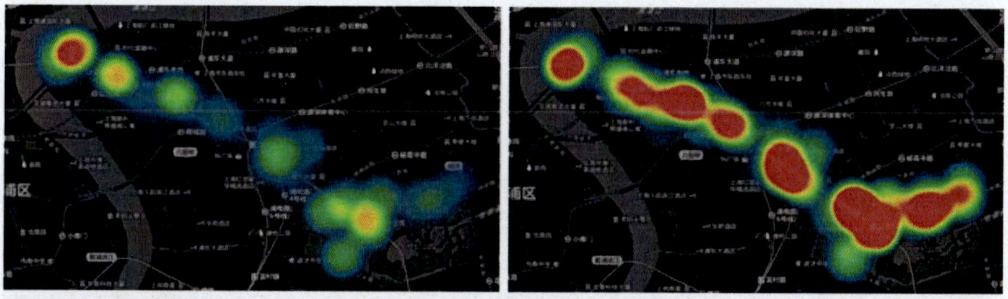

Fig. 4.33 Distribution of the number and dominance of Chinese Blackbird in the study area (left: number; right: dominance)

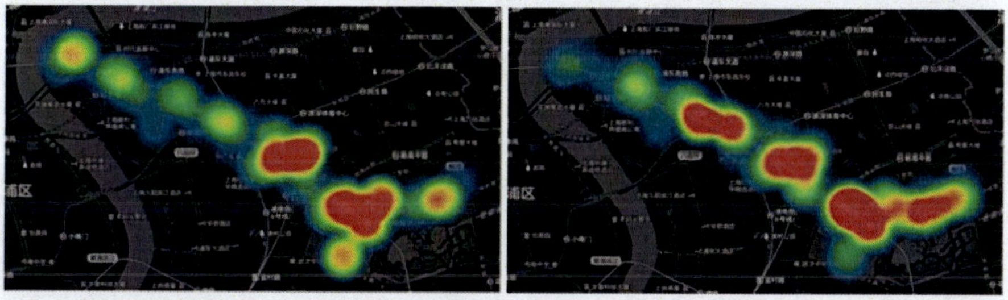

Fig. 4.34 Distribution of the number and dominance of Spotted Dove in the study area (left: number; right: dominance)

4.4.2 Aggregation by Preference

Azure-winged Magpie and White Wagtail, two species of insectivorous and arboreal birds, and Black-crowned Night Heron and Mew Gull, two species of piscivorous and waterside-dwelling birds, show the spatial distribution characteristic of high aggregation in the study area. All four species live in groups and have specific preferences for feeding and nesting niches. Black-crowned Night Heron and Mew Gull live in the Riverside area of Lujiazui Financial Center in large numbers, while Azure-winged

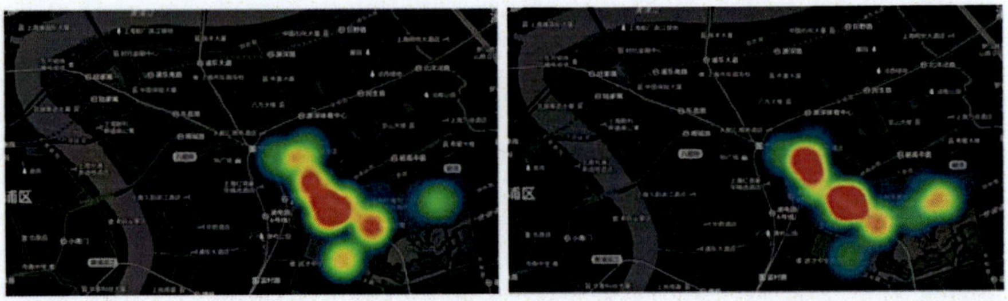

Fig. 4.35 Distribution of the number and dominance of Azure-winged Magpie in the study area (left: number; right: dominance)

Magpie, a dominant species in the Century Park, and White Wagtail, a common species, are highly aggregated in blocks close to the "source" habitat of the Century Park, forming stable populations (Figs. 4.35, 4.36, 4.37 and 4.38).

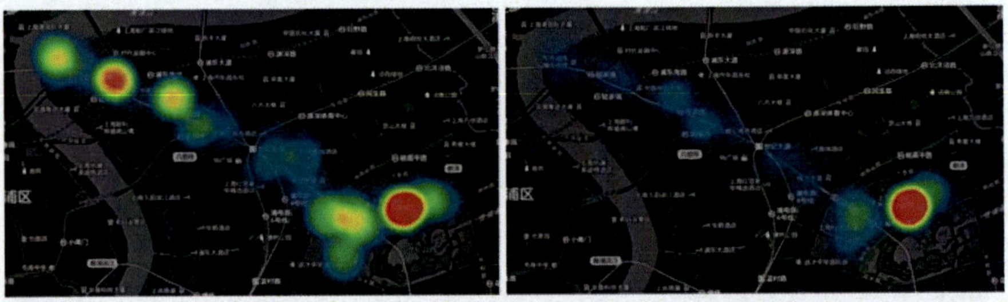

Fig. 4.36 Distribution of the number and dominance of White Wagtail in the study area (left: number; right: dominance)

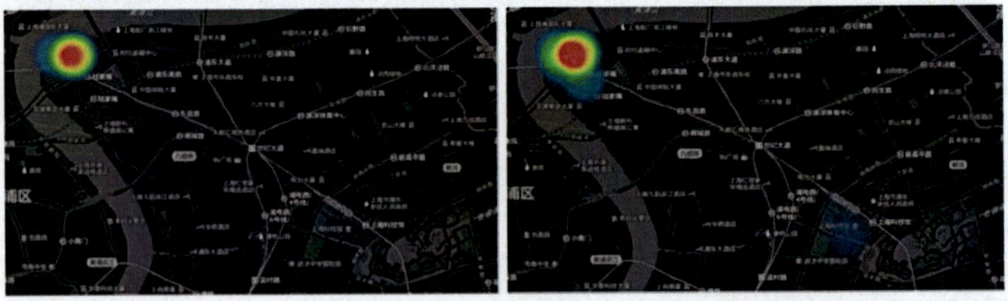

Fig. 4.37 Distribution of the number and dominance of Black-crowned Night Heron in the study area (left: number; right: dominance)

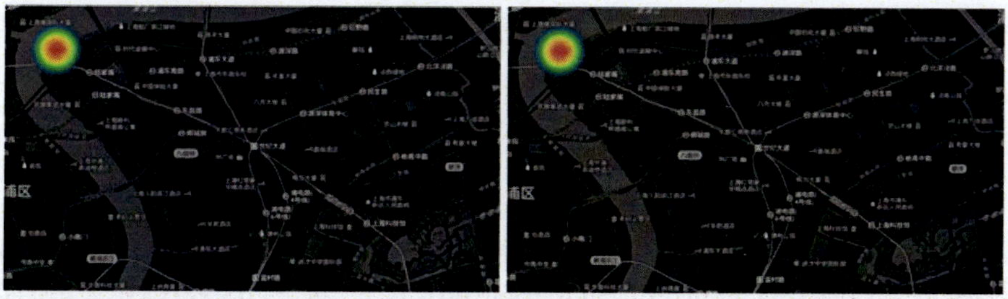

Fig. 4.38 Distribution of the number and dominance of Mew Gull in the study area (left: number; right: dominance)

4.4.3 Clustering at Scattered Points

The distribution of Vinous-throated Parrotbill and Long-tailed Shrike is relatively scattered. This is because they are more sensitive to the habitat environment and require larger habitat space to form communities, They are scattered in urban public green spaces, wetlands, and residential communities with better green environments (Figs. 4.39 and 4.40).

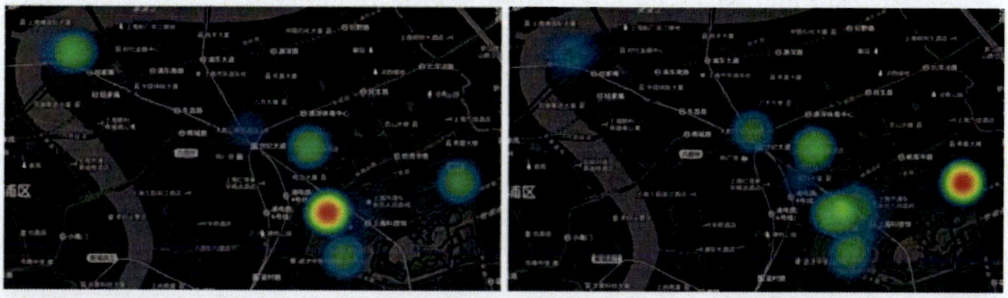

Fig. 4.39 Distribution of the number and dominance of Vinous-throated Parrotbill in the study area (left: number; right: dominance)

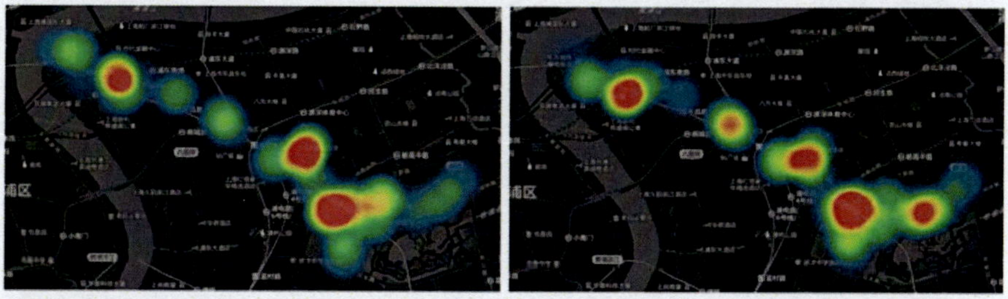

Fig. 4.40 Distribution of the number and index of Long-tailed Shrike in the study area (left: number; right: dominance)

4.4.4 Analysis of Preference for Proximity of Feeding and Nesting Niches

According to the research in Chap. 3, feeding niches and nesting niches of wildlife should be proximate to a certain extent. Whether they aggregate by preference or cluster at scattered points, when it comes to space selection, they have preferences for feeding niches, nesting niches, resting niches, etc. that are proximate to one another, that is, by centering around spaces that can provide stable food sources (e.g. fish and shrimp for water birds such as Black-crowned Night Heron) or nesting spaces (e.g. the Century Park with a large amount of thick woodland provides nesting spaces for Azure-winged Magpie) as "source" habitats, they expand appropriately, resulting in the characteristic of proximity clustering to a certain degree.

Take Black-crowned Night Heron as an example. Their feeding niches are riverside intertidal zones, whereas their nesting niches are shrubs or tree branches. Theoretically, the intertidal zone along the Huangpu River is the feeding niches of Black-crowned Night Heron. However, the author finds in the field survey that this is not necessarily true. Hundreds of Black-crowned Night Heron nest in the Oriental Pearl Park in Lujiazui. They feed on the intertidal zone by the Huangpu River 100 m away and fly back to their nests to feed their chicks, forming a stable population. In other green areas with tall trees suitable for nesting along the Huangpu River, however, there is hardly any Black-crowned Night Heron found (Fig. 4.41). Although the small wetland beside Shanghai Science and Technology Museum has a good nesting environment, the nearby Zhangjiabang water area is too artificial to be the feeding niches of Black-crowned Night Heron. Therefore, only a tiny number of passing

Fig. 4.41 The preference for proximity of Black-crowned Night Heron whose nesting niches are in the Oriental Pearl Park and feeding niches in the quagmire intertidal zone by the Huangpu River

Black-crowned Night Heron appear in this small wetland, and no nest of them was found. Given that, we need to consider the proximity effect of the two types of spatial niches in the spatial planning and design so that we can meet the needs of birds whose feeding and nesting niches do not completely overlap.

4.5 Spatial Distribution of Avian Diversity Variables

4.5.1 Overall Spatial Distribution of Avian Diversity Variables

In this study, the individual number of species, Species richness, biodiversity index, evenness index and superiority index are selected as the variables for evaluating avian biodiversity.

A total of 54 blocks were used as spatial statistical units, and the species diversity variables of each block were calculated separately and visualized with ArcGIS 10.2 software.

(1) **The spatial distribution of individual numbers of birds**

The block with the highest individual number of birds is A00, which contains the Oriental Pearl Park and the Oriental Pearl TV Tower. A total of 1,150 birds were monitored throughout the year, accounting for 11.24% of the total number. The individual numbers of birds in B17 (including the small wetland beside Shanghai Science and Technology Museum) and A02 (Lujiazui Central Greenland) are between 600 and 900. The individual numbers of birds in A13 (Zhuyuan New Village), AB00 (Riverside Greenland), B20 (Lujiazui Apartment), A21 (Lianyangxinyuan Community and Window of Shanghai·Yujingyuan Community), A06 (including Dongchang Middle School attached to East China Normal University) are between 300 and 600. Half of these blocks are large semi-natural habitats in the urban parks and public green spaces and habitats with natural-succession on the abandoned land, where there are wading bird species that breed in groups. For example, hundreds of Ardeidae birds gather in blocks A00 and AB00, which become blocks with more individual numbers of birds. The other half are residential areas and schools with better greening and semi-artificial habitats in the attached green spaces.

Based on the distribution of individual numbers of birds in all seasons, the agglomeration of birds in blocks with higher individual numbers is more obvious in spring and winter, and the distribution in summer and autumn is more balanced (Fig. 4.42).

(2) **Spatial distribution of species richness of birds**

In B17 block (the small wetland and surrounding woods beside Shanghai Science and Technology Museum) of the study area, there are up to 30 bird species, followed by B20 (Lujiazui Central Apartment, 23 species), A00 (including the Oriental Pearl Park, 22 species), B18 (square in front of the Shanghai Science and Technology

Fig. 4.42 Distribution of individual numbers of birds throughout the year and in four seasons in the study area

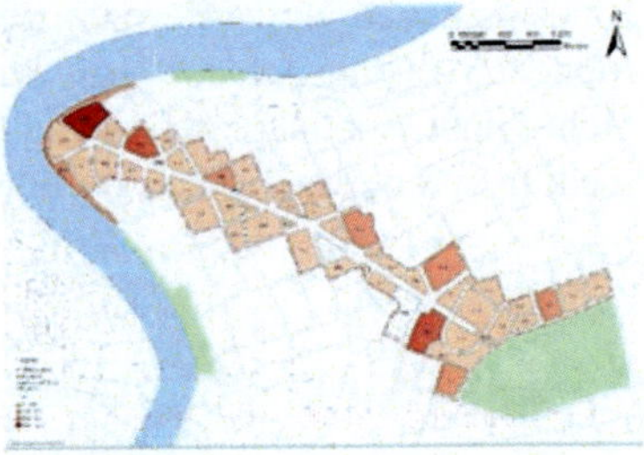

(a) All year

(b) Spring

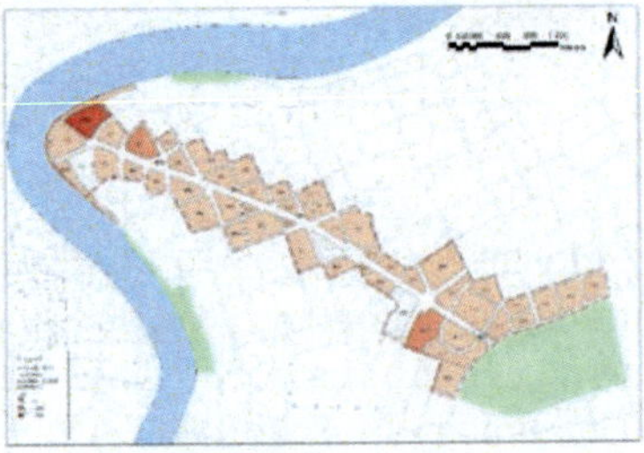

(c) Summer

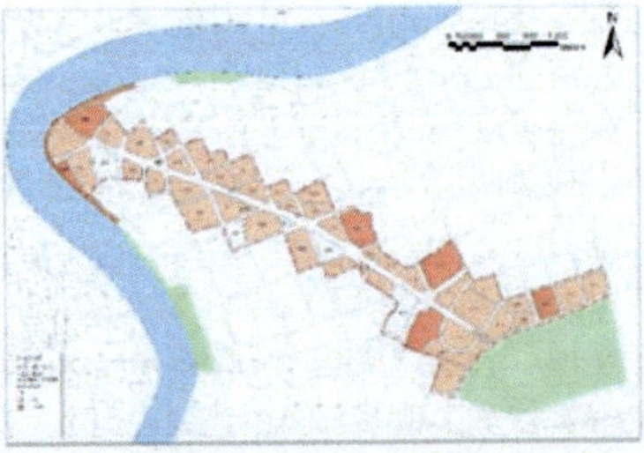

(d) Autumn

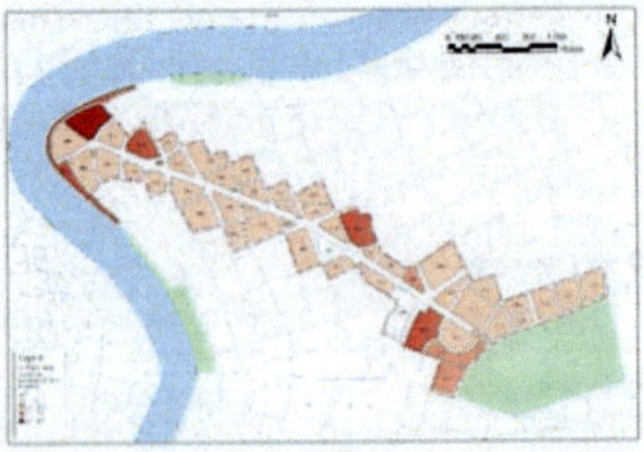

(e) Winter

Museum, 20 species), A18 (the Shanghai Science and Technology Museum Station Square, 18 species), AB00 (Riverside Greenland, 17 species), A21 (Lianyangxinyuan Community and Window of Shanghai·Yujingyuan Community, 17 species), B19 (Zhangjiabang Riverside Green Space behind the Shanghai Science and Technology Museum, 17 species), A02 (Lujiazui green space, 16 species). Overall speaking, the species richness is higher along the river and the Century Park, and lower in the residential areas and business areas in the middle section. Compared with the spatial distribution of individual numbers, it can be found that the peaks of the two figures have a certain correlation, but are not completely consistent. In other words, although there are more individual numbers of birds monitored in some blocks (such as A02 Lujiazui Central Greenland), their species richness is not high, and the bird species is relatively simple (Fig. 4.43).

As for the species richness in four seasons, the distribution in spring, autumn and winter is basically the same as that throughout the year, demonstrating high species richness along the Huangpu River and along the Century Park. However, the species richness along the Yangtze River significantly decreases in summer, which is mainly due to the northward migration of the Black-crowned Night Heron living in the riverside habitat in summer.

(3) Spatial distribution of diversity index of birds

The Shannon–Weiner diversity index of the study area ranges from 0 to 2.53. The blocks with an index value greater than 2.0 are B20 (Lujiazui Central Apartment), B17 (the small wetland and surrounding woods beside Shanghai Science and Technology Museum), A18 (the north of the Shanghai Science and Technology Museum Station Square), B18 (the south of the Shanghai Science and Technology Museum Station Square), and B19 (Zhangjiabang Riverside Green Space behind the Shanghai Science and Technology Museum), located along the Century Park. Blocks with a diversity index value between 1.5 and 2.0 are mainly located along the Century Park and along the Huangpu River. The distribution of the diversity index in spring, autumn and winter is basically the same as that in the whole year. In summer, the diversity index of the blocks along the Huangpu River and the Century Park drops significantly, but still higher than that of multi-storey residential areas and the Xinshanghai Commercial City area (Fig. 4.44).

(4) Spatial distribution for evenness index of birds

The bird evenness index (Pielou's evenness index) of the study area ranges from 0.228 to 1.000. The block with an evenness index value of 1.000 is B10 (Oriental Financial Plaza), but there are only two species, and the diversity index is only 0.693, which indicates that the numbers of birds of all species are roughly equal. The blocks with an evenness index value of 0.800–1.000 are A16 + (Douxiang Park), A20 (UnionPay Building + Jinying Plaza), A20 + (Pudong Expo Center), B14 (China Petroleum Shanghai Tower), and B15 (Orient Building +), located along the Century Park. Except for the Douxiang Park, the species richness and Shannon–Weiner diversity

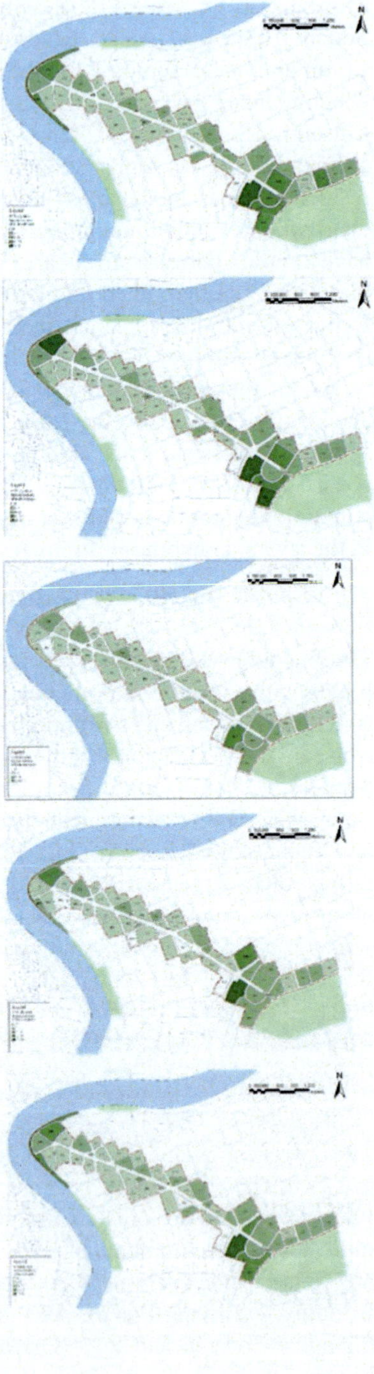

(a) All year

(b) Spring

(c) Summer

(d) Autumn

(e) Winter

Fig. 4.43 Distribution of species richness of birds throughout the year and in four seasons in the study area

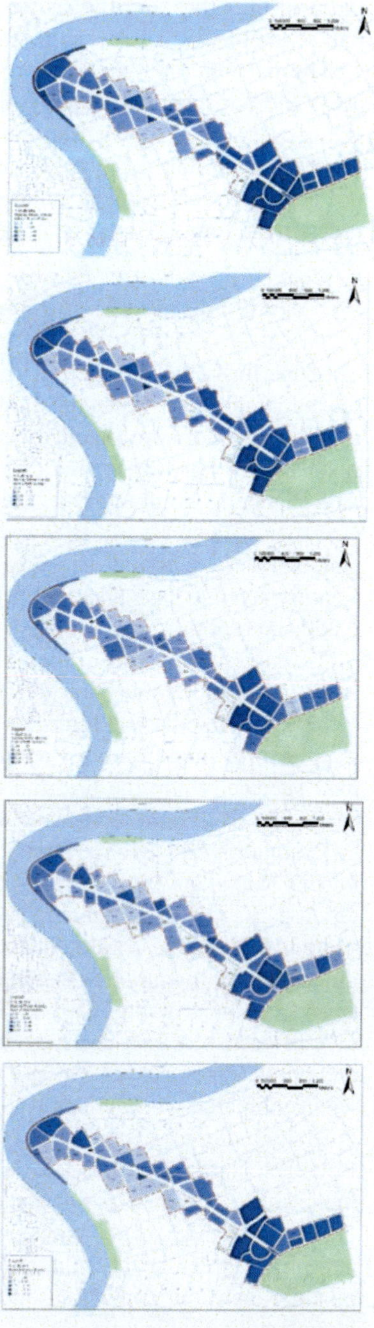

(a) All year

(b) Spring

(c) Summer

(d) Autumn

(e) Winter

Fig. 4.44 Distribution of bird diversity index in the study area throughout the year and in four seasons index of the other four blocks are not high. The evenness index value of more than half of the blocks is between 0.800 and 0.600.

The distribution of the evenness index of multi-storey commercial and residential areas and Xinshanghai Commercial City in the four seasons is quite different, that is, the evenness of the distribution of individual numbers of each bird species varies greatly. Given that the species richness and diversity index of these two areas do not change much in the four seasons, the main reason for the different evenness index in the four seasons is the seasonal fluctuation of the bird community structure (Fig. 4.45).

(5) Spatial distribution for superiority index of birds

The superiority index of birds in the study area ranges from 0.113 to 1.000. The larger the dominant index value, the more uneven the distribution of species in the community, and the more prominent the status of dominant species. Contrary to the species richness, diversity index, and evenness index, the superiority index is the highest in A05 (Sanhang Community in Meiyuan Subdistrict + Zhaoyuan Community), A07 (Laoshan Village No. 4), B09 and B09 + (Xindalu Square), which are multi-storey residential areas, commercial areas or business areas. Most of the blocks close to the Century Avenue have a superiority index value less than 0.200, and others between 0.200 and 0.600.

Through the analysis of the four seasons, it can also be found that the superiority index is higher in multi-storey residential areas, commercial and business areas, and lower in the middle and high-rise residential areas and green spaces. This indicates that the birds in multi-storey residential areas are mainly a limited number of dominant species, while the birds in middle and high-rise residential areas and green spaces are from more diverse sources (Fig. 4.46).

Comparing the avian diversity variables in the study area with the data collected by the Shanghai Wildlife Conservation Management Station in 2012 on 7 major urban parks and public green spaces including Gongqing National Forest Park, we can find that the mean value of species richness (9) of each block in the study area is much smaller than that of Daning Lingshi Park, Century Park, Shanghai Botanical Garden, and Gongqing National Forest Park. The highest value of species richness (47) of each block in the study area is also lower than that of other urban parks and public green spaces except Daning Lingshi Park, and the gap becomes greater when compared with the forest for water resources conservation in the suburbs. The diversity index of several blocks exceeds that of Daning Lingshi Park and Shanghai Botanical Garden, but overall, the average value of the diversity index of the blocks in the area is less than half of the total of the seven urban parks. The average value of the evenness index of the blocks in the area is close to the value of the 7 urban parks and public green spaces; the superiority index is too high, with an average value of more than four times the total value of the 7 urban parks, showing remarkable homogeneity (Table 4.4).

Fig. 4.45 Distribution of evenness index of birds throughout the year and in four seasons in the study area

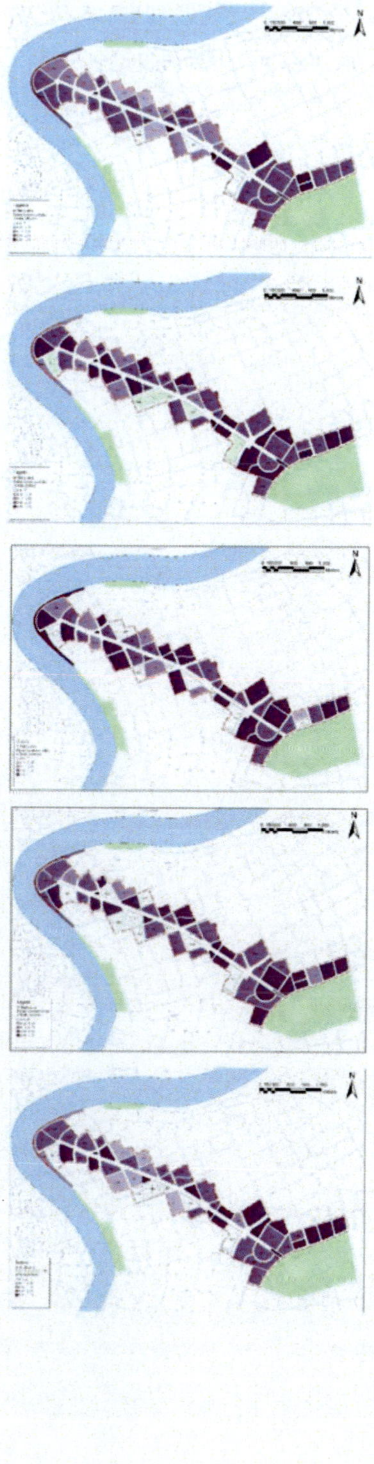

(a) All year

(b) Spring

(c) Summer

(d) Autumn

(e) Winter

Fig. 4.46 Distribution of superiority index of birds in the study area throughout the year and in four seasons

(a) All year

(b) Spring

(c) Summer

(d) Autumn

(e) Winter

Table 4.4 Comparison between the avian diversity variables of wild birds of the study area and of 7 urban parks and public green spaces in Shanghai

Variables	The study area	Gongqing National Forest Park	Daning Lingshi Park	Shanghai Botanical Garden	Century Park	Forest for water resources conservation in Luojing Town, Baoshan District	Shanghai Haiwan National Forest Park	Forest for water resources conservation in Punan Town, Songjiang District	7 urban parks in total
Species richness	0–47 (mean value 9)	64	44	58	55	88	83	63	136
Diversity index (Shannon-Weiner diversity index)	0–2.536 (mean value 1.359)	2.894	2.134	2.329	2.876	2.991	3.181	3.079	3.246
Evenness index (Pielou's evenness index)	0.228–1 (mean value 0.671)	0.696	0.564	0.573	0.718	0.668	0.720	0.743	0.661
Superiority index (Simpson's Diversity Index)	0.113–1 (mean value 0.373)	0.098	0.192	0.151	0.086	0.123	0.069	0.077	0.076

4.5.2 Distribution of Species Diversity Variables of Birds in Different Habitats

Through further analysis of the fluctuation range and average value of the avian diversity variables in the three different habitats and land-use types, we find that the species richness of birds is much higher in the habitats with natural-succession on the abandoned land than that in the other two types of habitats. Among semi-natural habitats in the urban parks and public green spaces, the protected green buffers have higher while roadside green spaces have lower species richness. According to the analysis in Sect. 4.2.4, the overall species richness is higher in roadside green spaces, but lower in a single block, indicating that the number of bird species that each roadside green space can attract is limited, but specific circumstances vary. This implies that the roadside green spaces in the study area serve as transit stepping stones for various birds in the city. In semi-artificial habitats in the attached green spaces, the species richness of the three types of residential land fluctuates greatly, and the average value is second only to that of the land for roads and squares with only two sampling points. The species richness of the land for business offices fluctuates the least, and its average value is only higher than that of the land for commercial services (Fig. 4.47).

The diversity index of birds in the habitats with natural-succession on the abandoned land is also much higher than that in the other two types of habitats. In semi-natural habitats in the urban parks and public green spaces, the fluctuation range of roadside green spaces is large, which once again shows the difference in species diversity. Among semi-artificial habitats in the attached green spaces, the fluctuation range of the diversity index of Land for Class III residential clusters, where there are mainly high-rise residential buildings, is between 1.3 and 2.3, significantly higher than that of land for Class II residential clusters, where there are mainly multi-storey residential buildings; and the average value is also higher than that of other land types, that is, the bird diversity in high-rise and low-density development areas is better than that in low-rise and high-density development areas. The fluctuation range and average value of the diversity index of the land for business offices are similar to those of land for Class II residential clusters (Fig. 4.48).

There is little difference in the evenness index of the three types of habitats. In semi-artificial habitats in the attached green spaces, except for the land for administration buildings (Pudong New Area People's Government) which has only 1 sampling point, and the land for education and research (Shanghai New York University), of which the evenness index is 1.000 and 0.859 respectively, the average value of other land types is between 0.6 and 0.8, and the fluctuation range of land for public facilities is greater than that of residential land (Fig. 4.49).

The superiority index of wetland habitats is the lowest, and below 0.4 of habitats in the urban parks and public green spaces. In semi-artificial habitats, the dominance on land for public facilities is greater than that of residential land (Fig. 4.50).

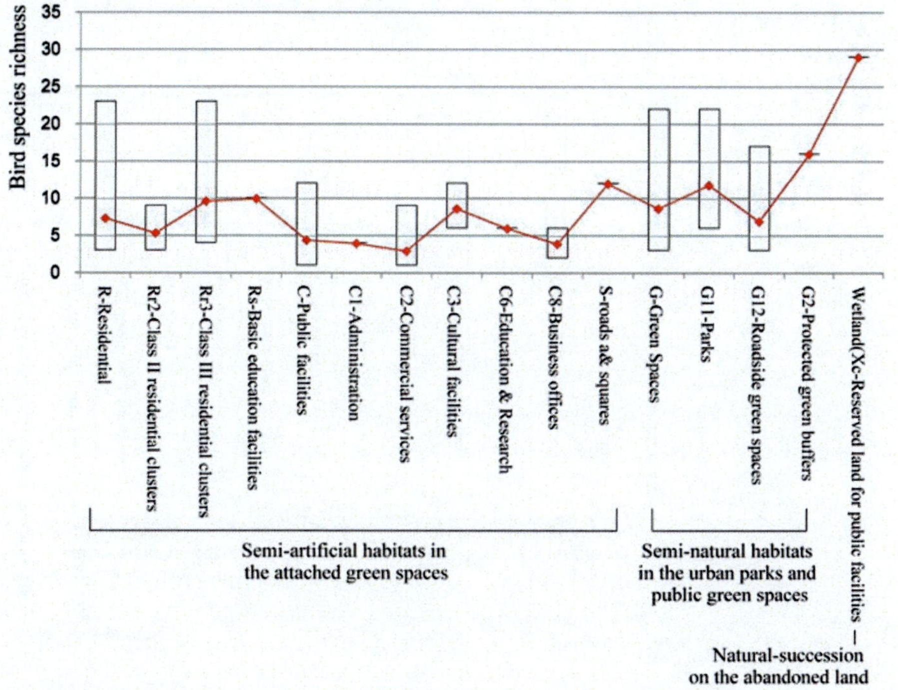

Fig. 4.47 Fluctuation range and average value of bird species richness of different habitats in the study area

4.6 Correlative Analysis Between Urban Built Environment and Avian Diversity

This book selects the built environment variables from three types: ecological land, vegetation and development intensity, specifically including:

(1) **Development intensity**

Development intensity variables include FAR, building density, green space ratio and water surface ratio; the first three are commonly used indicators in the regulatory plan. As water bodies are regarded as niches providing drinking water for birds and have a certain attraction to birds, water surface ratio is also analyzed.

(2) **Ecological land**

Ecological land variables include the ecological land area and spatial form variables. Considering the operability of planning optimization, this study refers to spatial pattern metrics in landscape ecology as the characteristic variables, including total class area, perimeter area ratio, patch density, largest patch index, largest patch area, largest patch perimeter area ratio, mean proximity index (MPI), patch aggregation

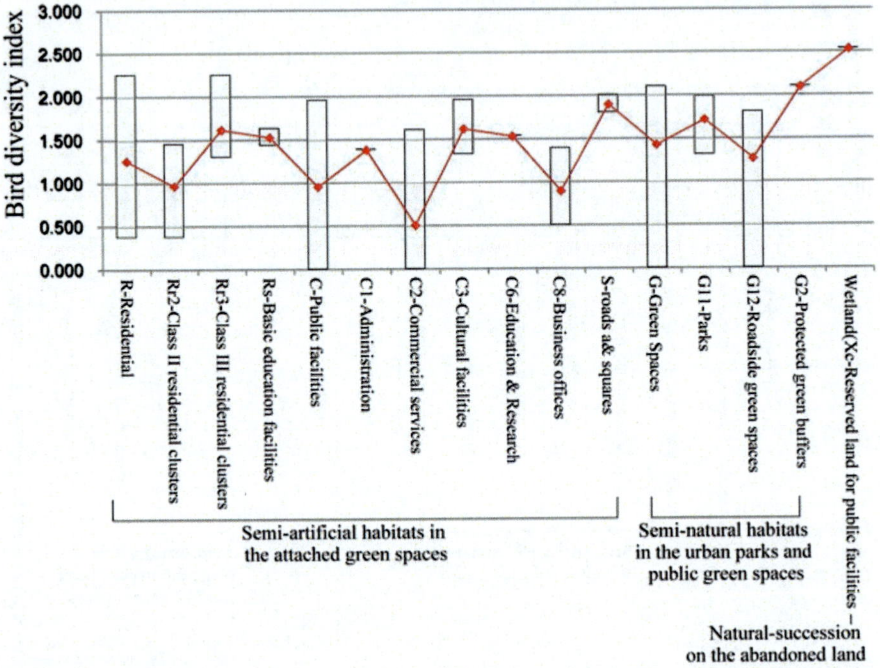

Fig. 4.48 The fluctuation range and mean value of bird diversity index of different habitats in the study area

index (AI) and Patch Cohesion Index of green space and water body. As only 1/3 of the 51 blocks have water bodies, and generally there are one or two complete water surfaces in one block, it is impossible to calculate its fragmentation degree and connectivity. Therefore, the largest patch perimeter area ratio, mean proximity index, patch aggregation index and patch cohesion index are calculated only for green space patches.

(3) **Vegetation variables**

Vegetation variables used in this research include the coverage area, height and species richness of trees, shrubs and groundcover plants (Table 4.5).

4.6.1 Correlation Analysis Between Development Intensity and Avian Diversity

The analysis results show that FAR and building density is significantly negatively correlated with species richness and diversity index (p-value = 0.01), while significantly positively correlated with superiority index (p-value = 0.01). The correlation

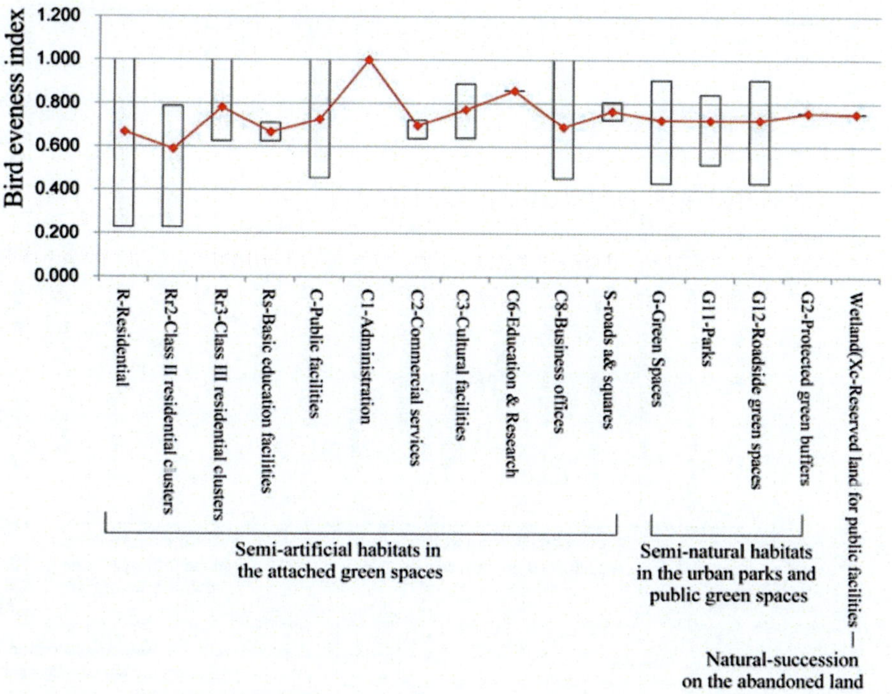

Fig. 4.49 The fluctuation range and mean value of bird evenness index of different habitats in the study area

coefficients between building density with the avian diversity variables are mostly higher than those between FAR with the avian diversity variables, which mean, the higher the building density, the more space occupied by human activities, and the fewer ground space left for wild birds.

The green space ratio is positively correlated with species richness and diversity (p-value = 0.01), as well as individual number of birds (p-value = 0.05), which indicates that a higher proportion of green space area is beneficial to support more habitat space for birds. There is a significant positive correlation between species richness, diversity index and water surface ratio (p-value = 0.01), but the correlation coefficients are smaller. This may be due to the fact that most of the water surface in the study area is man-made landscape, which can only provide certain water source for birds, but not food for piscivorous. Moreover, some water bodies in residential areas will dry up due to poor management. Therefore, the water surfaces might have a certain attraction to birds, but the impact is relatively smaller compared with green space (Table 4.6).

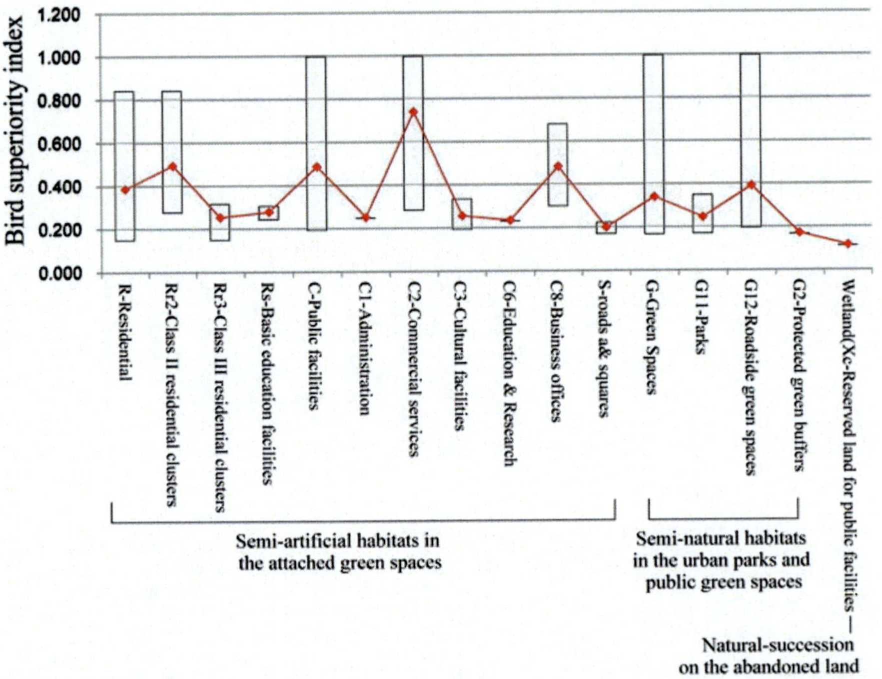

Fig. 4.50 The fluctuation range and mean value of bird superiority index of different habitats in the study area

4.6.2 *Correlation Analysis Between Ecological Land and Avian Diversity*

The research results show that the class area and largest patch area of green space are significantly positively correlated with the individual number, species richness and diversity index of birds(p-value = 0.01), and significantly negatively correlated with superiority index(p-value = 0.01). The patch density of green space has significantly negative correlation with species richness (p-value = 0.05), evenness index (p-value = 0.05), diversity index (p-value = 0.01), while is significantly positively correlated with superiority index (p-value = 0.01), which verified the theory of "species-area relation" of the conservation ecology.

The perimeter area ratio of total and of the largest patch of green space are significantly negatively correlated with individual number, species richness and diversity index (p-value = 0.01), and is positively correlated with superiority index (p-value = 0.01), which indicates that the disturbance by edge effect has a great negative influence on the biodiversity at the medium-micro scale.

There is a significant negative correlation between the largest patch index of green space and evenness index (p-value = 0.01), which indicates that a concentrated green space with a larger area has a better positive effect on the avian diversity.

Table 4.5 Selection of vegetation variables

Variable type	Variable name	Variable unit
Vegetation area	Tree coverage area	m^2
	Shrub coverage area	m^2
	Coverage area of groundcover plants	m^2
	Percentage of land covered by trees	%
	Percentage of land covered by shrubs	%
	Percentage of land covered by groundcover plants	%
Vegetation form	Maximum height of trees	m
	Average height of trees	m
	Average height of shrubs	m
	Maximum height of shrubs	m
Vegetation structure	Species richness of trees	–
	Species richness of evergreen trees	–
	Species richness of deciduous trees	
	Species richness of shrubs	–
	Species richness of evergreen shrubs	–
	Species richness of deciduous shrubs	–
	Species richness of groundcover plants	–
	Species richness of plants	–

Table 4.6 Pearson's correlation analysis between development intensity variables and avian diversity variables

	Individual number	Species richness	Diversity index	Evenness index	Superiority index
FAR	**−0.437****	**−0.520****	**−0.456****	0.144	**0.339****
Building density (%)	**−0.397****	**−0.571****	**−0.684****	**−0.359***	**0.655****
Green space ratio (%)	**0.355***	**0.517****	**0.562****	0.134	**−0.514****
Water surface ratio (%)	0.207	**0.341***	**0.311***	0.109	−0.231

Note *Correlation is significant at a P value of 0.05 (two-sided test). **Correlation is significant at a P value of 0.01 (two-sided test)

Mean proximity index, patch aggregation index and patch cohesion index of green space are positively correlated with individual number, species richness, diversity index and evenness index, while negatively correlated with superiority index, which reflects the distance effect: the closer the distance between patches, the better the connectivity, the diverse bird species would be attracted, and the higher the avian diversity.

There is also a significant positive correlation between the total class area, patch average area of water and bird species richness (p-value = 0.05). Perimeter area ratio of water is significantly positively correlated with diversity index (p-value = 0.05) and evenness index (p-value = 0.01), and is negatively correlated with superiority index (p-value = 0.01). Largest patch index of water is significantly positively correlated with diversity index (p-value = 0.05) and evenness index (p-value = 0.01), and is significantly negatively correlated with superiority index (p-value = 0.01).

There is a significant positive correlation between the species richness (p-value = 0.01) and diversity index (p-value = 0.05), while negatively correlated with superiority index (p-value = 0.01), which indicates that the water body has a certain impact on the avian diversity (Table 4.7).

4.6.3 Correlation Analysis Between Vegetation and Avian Diversity

The results show that coverage area of trees, shrubs and groundcover plants are significantly positively correlated with individual number, species richness and diversity index of bird, while negatively correlated with superiority index. Percentage of land covered by trees is significantly positively correlated with diversity index (p-value = 0.01), and negatively correlated with superiority index (p-value = 0.05). Percentage of land covered by shrubs is significantly positively correlated with evenness index (p-value = 0.05). Percentage of land covered by groundcover plants is significantly positively correlated with diversity index (p-value = 0.01), and negatively correlated with superiority index (p-value = 0.01).

Comparatively speaking, the correlation coefficient between avian diversity variables with coverage area of trees is relatively higher than that of shrubs and groundcover plants; while the correlation coefficient between avian diversity variables with percentage of land covered groundcover plants are higher than the other two types of vegetation. This indicates that effects of tree layer and groundcover layer on avian diversity are greater than that of the shrub layer.

The correlation analysis results between the vegetation height and the avian diversity show that the average height of tree is significantly positively correlated with species richness and diversity index (p-value = 0.01), individual number (p-value = 0.05), and negatively correlated with superiority index (p-value = 0.05). The maximum height of trees is positively correlated with individual number and species richness (p-value = 0.01), diversity index (p-value = 0.05), and negatively correlated with the superiority index (p-value = 0.05). Also there is a significant positive correlation between the average and the maximum height of shrubs with individual number (p-value = 0.05). This represents that tree height, especially the average height, has a larger influence on the avian diversity, which reflects that the niche extending to the vertical space might provide a broader off-ground habitat for wild birds.

Table 4.7 Pearson's correlation analysis between ecological land variables and avian diversity variables

	Individual number	Species richness	Diversity index	Evenness index	Superiority index
Total class area of green space (m^2)	**0.772****	**0.803****	**0.595****	−0.042	**−0.454****
Patch density of green space (N/m^2)	−0.236	**−0.322***	**−0.518****	**−0.296***	**0.593****
Perimeter area ratio of green space (m/m^2)	**−0.356****	**−0.496****	**−0.646****	**−0.311***	**0.681****
Largest patch index of green space (%)	−0.083	−0.019	0.146	**0.449****	−0.243
Largest patch area of green space (m^2)	**0.640****	**0.690****	**0.515****	0.047	**−0.402****
Perimeter area ratio of the largest patch of green space (m/m^2)	**−0.377****	**−0.490****	**−0.581****	−0.122	**0.617****
Mean proximity index of green space	**0.295***	**0.368****	**0.342***	0.075	**−0.295***
Patch aggregation index of green space	**0.371****	**0.508****	**0.666****	**0.308***	**−0.708****
Patch cohesion index of green space	**0.340***	**0.490****	**0.670****	**0.328***	**−0.697****
Total class area of water (m^2)	0.247	**0.303***	0.274	0.050	−0.219
Patch average area of water (m^2)	0.234	**0.275***	0.214	−0.008	−0.162
Perimeter area ratio of water (m/m^2)	0.056	0.205	**0.372****	**0.322***	**−0.370****
Largest patch index of water (%)	0.027	**0.313***	**0.386****	0.200	**−0.330***

Note *Correlation is significant at a P value of 0.05 (two-sided test). **Correlation is significant at a P value of 0.01 (two-sided test)

Species richness of vegetation is positively correlated with individual number and species richness (p-value = 0.01), diversity index (p-value = 0.05), and negatively correlated with superiority index (p-value = 0.05). Generally speaking, the correlation coefficients between avian diversity variables and species richness of groundcover plants is higher than those of trees, and those of shrubs are the lowest. Combined with higher correlation coefficients between the proportion of land covered by groundcover plant and avian diversity variables above, this reflects the importance of groundcover planting, especially for birds feeding on the ground, which also provides more feeding and rest space (Table 4.8).

4.7 Impact of the Built Environment of Various Functional Areas on Avian Diversity

According to the land-use type of more than 50% of the area of a block, the 51 blocks are divided into three types of functional areas, namely, blocks for public green spaces, residential blocks, and blocks for public facilities (land for commercial services, land for business offices, land for cultural facilities, etc.), and there are 8 blocks of public green spaces, 15 blocks of residential area and 22 blocks of public facilities. After separate analysis of the impact of the built environment of the three types of functional areas on avian diversity, and the results are shown in Sects. 4.7.1, 4.7.2 and 4.7.3.

4.7.1 Blocks for Public Green Spaces

Blocks for public green spaces that represent semi-natural habitats include the Oriental Pearl Park, Lujiazui Central Greenland, Douxiang Park, Riverside Greenland, Century Plaza-Century Park linear green corridor, green space in the administrative center of Pudong New Area, nursery at Jinxiu Road Metro Station and 1 traffic island green space.

The analysis results show that there is no significant correlation between the development intensity variables of blocks for public green spaces and avian diversity variables, while in the correlation analysis with the ecological land variables, only the total class area, patch density, largest patch area of green space and perimeter area ratio of green space show significant correlations with species richness(p-value = 0.05), and various variables of water patches show no significant correlations. It can be inferred that in public green spaces, large-area artificial water bodies have little effect on avian diversity.

Among the vegetation variables, the height and species richness of trees have a higher correlation coefficient with species diversity. Different from the results of overall correlation analysis, in blocks for public green spaces, shrubs (deciduous

Table 4.8 Pearson's correlation analysis between vegetation variables and avian diversity variables

	Individual number	Species richness	Diversity index	Evenness index	Superiority index
Tree coverage area (m^2)	**0.503****	**0.656****	**0.516****	0.092	**−0.386****
Shrub coverage area (m^2)	**0.350***	**0.457****	**0.366****	0.151	**−0.313***
Coverage area of ground cover (m^2)	**0.415****	**0.391****	**0.388****	0.123	**−0.320***
Percentage of land covered by trees (%)	0.046	0.234	**0.355***	0.234	**−0.335***
Percentage of land covered by shrubs (%)	−0.118	0.035	0.189	**0.300***	−0.242
Percentage of land covered by ground cover (%)	0.095	0.119	**0.392****	0.157	**−0.366****
Maximum height of trees (m)	**0.453****	**0.445****	**0.319***	0.074	**−0.281***
Average height of trees (m)	**0.286***	**0.480****	**0.520****	**0.295***	**−0.426****
Average height of shrubs (m)	**0.321***	0.256	0.178	−0.012	−0.132
Maximum height of shrubs (m)	**0.299***	0.230	0.117	−0.005	−0.072
Species richness of trees	**0.509****	**0.386****	0.140	−0.192	−0.104
Species richness of evergreen trees	**0.515****	**0.404****	0.162	−0.133	−0.138
Species richness of deciduous trees	**0.462****	**0.342***	0.114	−0.207	−0.075
Species richness of shrubs	**0.364****	**0.430****	**0.351***	0.075	**−0.311***
Species richness of evergreen shrubs	**0.328***	**0.444****	**0.383****	0.125	**−0.335***
Species richness of deciduous shrubs	**0.274***	0.215	0.113	−0.058	−0.109
Species richness of groundcover plants	**0.536****	**0.657****	**0.531****	0.176	**−0.440****

(continued)

Table 4.8 (continued)

	Individual number	Species richness	Diversity index	Evenness index	Superiority index
Species richness of plants	**0.545****	**0.529****	**0.342***	−0.038	**−0.284***

Note *Correlation is significant at a P value of 0.05 (two-sided test). **Correlation is significant at a P value of 0.01 (two-sided test)

shrub species) show a significant and strong correlation with avian diversity, while the impact of groundcover plants greatly declines. This may be because most public green spaces have more non-sealed bare ground, and the effect of ground cover plants does not apply to blocks for public green spaces (Table 4.9).

4.7.2 Residential Blocks

Residential blocks boast the highest bird species richness among all types of blocks except public green spaces. The analysis results show that among the development intensity variables of residential blocks, the negative effect of FAR is not significant, while the building density, green space ratio and water surface ratio all show high correlation coefficients. This indicates that if high-rise and low-density residential areas allocate some ground habitat space to maintain a high ratio of green space and water surface, they may have higher avian diversity than low-rise and high-density residential areas. The correlation between the landscape metrics variables of green space patches and the avian diversity variables is similar to the overall level, while the largest water patch area shows a high correlation in the residential area, indicating that increasing the water body patch area in a residential area can effectively enhance species diversity. Among the vegetation variables, the correlation coefficients of coverage area, height, and species richness of shrubs increase significantly, while the impact of groundcover plants decrease slightly compared with the overall situation, indicating that reasonable distribution of the three vegetation layers of trees, shrubs and groundcover grasses in a residential area greatly enhances species diversity (Table 4.10).

4.7.3 Blocks for Public Facilities

Land for public facilities is a typical type of land in semi-artificial habitats in the attached green spaces in center of a high-density city. The analysis results show that the three development intensity variables of blocks for public facilities, namely,

Table 4.9 Pearson's correlation analysis between built environment variables of blocks for public green spaces in the study area and avian diversity variables

Variable type	Variable name	Individual number	Species richness	Diversity index	Evenness index	Superiority index
Development intensity	Building density (%)	0.612	0.218	−0.082	−0.444	0.137
	FAR	0.683	0.315	−0.154	−0.556	0.261
	Green space ratio (%)	−0.550	−0.246	−0.103	0.110	−0.006
	Water surface ratio (%)	0.254	0.561	0.464	0.115	−0.261
Ecological land	Total class area of green space (m^2)	0.672	**0.832***	0.563	−0.352	−0.346
	Perimeter area ratio of green space (m/m^2)	−0.355	−0.668	−0.481	0.305	0.340
	Patch density of green space (N/m^2)	−0.497	−0.646	−0.536	0.287	0.488
	Largest patch index of green space (%)	−0.436	−0.164	0.069	0.378	−0.150
	Largest patch area of green space (m^2)	0.493	**0.784***	0.605	−0.160	−0.394
	Perimeter area ratio of the largest patch of green space(m/m^2)	−0.690	**−0.809***	−0.488	0.481	0.337
	Mean proximity index of green space	−0.052	0.108	0.106	−0.156	−0.102
	Patch aggregation index of green space	0.173	0.467	0.254	−0.274	−0.123
	Patch cohesion index of green space	0.210	0.576	0.428	−0.228	−0.303

(continued)

Table 4.9 (continued)

Variable type	Variable name	Individual number	Species richness	Diversity index	Evenness index	Superiority index
	Total class area of water (m^2)	0.176	0.184	−0.105	−0.397	0.131
	Perimeter area ratio of water (m/m^2)	−0.568	−0.482	−0.169	0.419	0.081
	Patch average area of water (m^2)	0.205	−0.023	−0.421	−0.757	0.470
	Largest patch index of water (%)	0.684	0.599	−0.251	−0.759	0.392
	Largest water patch area (m^2)	0.199	−0.020	−0.410	−0.747	0.459
Vegetation	Tree coverage area (m^2)	0.416	0.440	0.068	−0.580	0.059
	Shrub coverage area (m^2)	0.357	0.066	−0.441	**−0.799***	0.533
	Coverage area of ground cover (m^2)	0.035	0.013	0.431	0.431	−0.509
	Percentage of land covered by trees (%)	−0.601	−0.459	−0.180	0.307	−0.022
	Percentage of land covered by shrubs (%)	−0.609	−0.551	−0.298	0.318	0.102
	Percentage of land covered by ground cover (%)	−0.513	−0.352	0.353	**0.733***	−0.607
	Maximum height of trees (m)	**0.725***	**0.793***	0.346	−0.530	−0.107
	Average height of trees (m)	0.706	**0.904****	0.686	−0.088	−0.444
	Average height of shrubs (m)	0.257	0.277	−0.182	−0.437	0.414

(continued)

Table 4.9 (continued)

Variable type	Variable name	Individual number	Species richness	Diversity index	Evenness index	Superiority index
	Maximum height of shrubs (m)	0.442	0.287	−0.263	−0.657	0.491
	Species richness of trees	0.547	0.264	−0.268	**−0.838****	0.365
	Species richness of evergreen trees	0.624	0.331	−0.265	**−0.877****	0.395
	Species richness of deciduous trees	0.466	0.198	−0.267	**−0.784***	0.330
	Species richness of shrubs	0.283	0.164	−0.154	−0.497	0.218
	Species richness of evergreen shrubs	0.275	0.272	0.067	−0.319	0.002
	Species richness of deciduous shrubs	0.623	**0.760***	−0.170	**−0.803***	0.375
	Species richness of groundcover plants	0.444	0.613	0.263	−0.505	−0.140
	Species richness of plants	0.493	0.337	−0.139	**−0.736***	0.240

Note *Correlation is significant at a P value of 0.05 (two-sided test). **Correlation is significant at a P value of 0.01 (two-sided test)

building density, FAR, and green space ratio, are significantly correlated to the bird species richness. The correlation between the landscape metrics variables of green space patches and the avian diversity variables is similar to the overall situation, while the total class area of water, perimeter area ratio, average area and the largest patch index show correlation coefficient values ranging from 0.4 to 0.6, with many data showing significant correlation. In fact, although the water bodies in blocks for public facilities are generally small and scattered, they still have a certain effect, which indicates that increasing water body patch area in land for public facilities such as

Table 4.10 Pearson's correlation analysis between built environment variables of residential blocks in the study area and avian diversity variables

Variable type	Variable name	Individual number	Species richness	Diversity index	Evenness index	Superiority index
Development intensity	Building density (%)	−0.394	**−0.649****	**−0.886****	**−0.710****	**0.884****
	FAR	−0.193	−0.036	−0.050	0.063	0.078
	Green space ratio (%)	0.193	0.342	**0.614***	**0.554***	**−0.631***
	Water surface ratio (%)	0.080	**0.600***	**0.682****	0.439	**−0.522***
Ecological land	Total class area of green space (m^2)	**0.621***	**0.632***	**0.557***	0.224	−0.456
	Perimeter area ratio of green space (m/m^2)	−0.397	−0.422	**−0.646****	**−0.581***	**0.705****
	Patch density of green space (N/m^2)	−0.413	**−0.546***	**−0.684****	**−0.514***	**0.669****
	Largest patch index of green space (%)	0.137	0.050	0.240	0.403	−0.356
	Largest patch area of green space (m^2)	**0.577***	0.506	**0.577***	0.445	**−0.582***
	Perimeter area ratio of the largest patch of green space(m/m^2)	−0.302	−0.301	−0.408	−0.321	0.449
	Mean proximity index of green space	0.087	0.446	**0.519***	−0.473	−0.361
	Patch aggregation index of green space	0.277	0.409	**0.669****	**0.615***	**−0.705****
	Patch cohesion index of green space	0.338	0.404	**0.679****	**0.636***	**−0.750****

(continued)

Table 4.10 (continued)

Variable type	Variable name	Individual number	Species richness	Diversity index	Evenness index	Superiority index
	Total class area of water (m^2)	−0.191	0.134	0.383	0.375	−0.356
	Perimeter area ratio of water (m/m^2)	−0.192	0.130	0.440	0.440	−0.405
	Patch average area of water (m^2)	0.154	0.368	0.311	0.109	−0.228
	Largest patch index of water (%)	0.134	0.372	0.362	0.166	−0.276
	Largest water patch area (m^2)	0.233	**0.714****	**0.651****	0.316	−0.466
Vegetation	Tree coverage area (m^2)	0.335	**0.701****	**0.535***	0.196	−0.345
	Shrub coverage area (m^2)	0.356	**0.728****	**0.750****	0.468	**−0.619***
	Coverage area of ground cover (m^2)	**0.580***	0.298	0.235	0.104	−0.281
	Percentage of land covered by trees (%)	0.216	0.503	0.471	0.270	−0.372
	Percentage of land covered by shrubs (%)	0.227	0.502	**0.626***	0.469	**−0.577***
	Percentage of land covered by ground cover (%)	0.215	0.041	0.144	0.200	−0.251
	Maximum height of trees (m)	0.078	0.114	0.105	0.196	−0.122
	Average height of trees (m)	0.051	0.357	0.411	0.248	−0.305

(continued)

Table 4.10 (continued)

Variable type	Variable name	Individual number	Species richness	Diversity index	Evenness index	Superiority index
	Average height of shrubs (m)	0.246	0.495	**0.591***	0.439	**−0.518***
	Maximum height of shrubs (m)	0.307	0.470	0.450	0.226	−0.421
	Species richness of trees	0.124	0.122	−0.091	−0.236	0.141
	Species richness of evergreen trees	0.296	0.482	0.455	0.184	−0.405
	Species richness of deciduous trees	0.017	−0.068	−0.314	−0.366	0.351
	Species richness of shrubs	0.419	**0.722****	**0.793****	**0.497**	**−0.709****
	Species richness of evergreen shrubs	0.483	**0.729****	**0.759****	**0.403**	**−0.679****
	Species richness of deciduous shrubs	−0.118	0.025	0.119	0.235	−0.106
	Species richness of groundcover plants	0.336	**0.661****	**0.780****	**0.609***	**−0.709****
	Species richness of plants	0.391	**0.640***	**0.575***	0.281	−0.479

Note *Correlation is significant at a P value of 0.05 (two-sided test). **Correlation is significant at a P value of 0.01 (two-sided test)

commercial areas and business areas can effectively enhance avian diversity. Among the vegetation variables, the influential effects of trees and ground cover are greater. Among the variables of the shrubs, except that the shrub coverage area is significantly correlated to the number and species richness of birds, and the percentage of land covered by shrubs is significantly correlated to the evenness index, other variables show no significant correlations. The reason may be that there are more interferences from human activities in blocks for public facilities. Birds mainly use the resting and feeding niches of the tree canopy layer and the feeding niches of the ground cover, while shrubs mainly serves as partition hedge (Table 4.11).

4.8 Influence Mechanism of Built Environment on Avian Diversity in Roadside Green Spaces

4.8.1 Selection of Sample Blocks for Study in Roadside Green Spaces

Roadside green spaces are small green space patch units with a certain scale that are easier to observe in the research, and have the potential to serve as "stepping stone" habitats. Sections 4.2.4 and 4.5.2 proves that the status of avian diversity in roadside green spaces in the overall green space habitat entails further examination. The Century Avenue is unique in that a series of triangular green spaces have taken shape at the irregular intersections. Most of them are marked as "temporary green spaces" in the greening management, but they received better artificial vegetation for the landscape effect when the Century Avenue was constructed, and have existed for more than 15 years. These green areas are maintained and pruned regularly by the greening management department, and have also experienced vegetation succession. After site surveys, it is found that there are 13 triangular roadside green spaces along the Century Avenue. According to the spatial relationship of these triangular green spaces with the Century Avenue and surrounding blocks, they can be divided into traffic island green spaces, road's end green spaces and embedded roadside green spaces.

This study divides the borders of the triangular roadside green spaces into interfered borders and proximate borders. Interfered borders refer to the borders adjacent to sidewalks or roadways that are interfered by outside human activities; and proximate borders refer to the borders adjacent to urban functional groups such as proximate residential areas, business areas, and commercial areas. Although clear separations and even walls are used to separate roadside green spaces from the proximate functional groups on the land boundaries, the canopies of their trees are generally higher than walls, connecting with the tree canopies of their proximate residential areas. Therefore, this kind of border can be considered to be friendly to species diversity.

The three borders of a traffic island green space are all interfered borders. Interfered borders of a road's end green space are longer than proximate borders, while interfered borders of an embedded roadside green space are shorter than proximate borders (Fig. 4.51).

Table 4.11 Pearson's correlation analysis between built environment variables of blocks for public facilities in the study area and avian diversity variables

Variable type	Variable name	Individual number	Species richness	Diversity index	Evenness index	Superiority index
Development intensity	Building density (%)	−0.293	**−0.570****	**−0.746****	**−0.568****	**0.753****
	FAR	**−0.536***	**−0.648****	**−0.527***	0.002	0.348
	Green space ratio (%)	0.402	**0.636****	**0.640****	0.235	**−0.581****
	Water surface ratio (%)	0.051	0.100	0.232	0.289	−0.288
Ecological land	Total class area of green space (m^2)	**0.687****	**0.745****	**0.564****	0.074	**−0.426***
	Perimeter area ratio of green space (m/m^2)	−0.327	**−0.482***	**−0.624****	**−0.693****	**0.702****
	Patch density of green space (N/m^2)	−0.231	−0.319	**−0.504***	**−0.746****	**0.614****
	Largest patch index of green space (%)	−0.179	−0.093	0.021	0.410	−0.187
	Largest patch area of green space (m^2)	0.421	**0.544****	**0.436***	0.064	−0.392
	Perimeter area ratio of the largest patch of green space(m/m^2)	−0.324	−0.420	**−0.571****	**−0.631****	**0.635****
	Mean proximity index of green space	0.310	**0.495***	−0.386	**0.594****	0.148
	Patch aggregation index of green space	0.359	**0.663****	**−0.722****	**0.512***	**0.722****
	Patch cohesion index of green space	0.344	**0.656****	**−0.711****	**0.510***	**0.682****

(continued)

Table 4.11 (continued)

Variable type	Variable name	Individual number	Species richness	Diversity index	Evenness index	Superiority index
	Total class area of water (m^2)	0.368	0.369	**0.438***	0.276	−0.430*
	Perimeter area ratio of water (m/m^2)	0.360	0.384	**0.522***	0.286	−0.499*
	Patch average area of water (m^2)	**0.612****	**0.522***	**0.428***	0.133	−0.382
	Largest patch index of water (%)	**0.495***	**0.530***	**0.510***	0.170	**−0.444***
	Largest water patch area (m^2)	**0.567****	**0.541****	**0.436***	0.138	−0.373
Vegetation	Tree coverage area (m^2)	**0.515***	**0.658****	**0.483***	0.077	−0.345
	Shrub coverage area (m^2)	**0.537****	**0.490***	0.318	0.047	−0.236
	Ground cover coverage area (m^2)	**0.609****	**0.700****	**0.468***	−0.009	−0.373
	Percentage of land covered by trees (%)	0.322	**0.565****	**0.566****	0.329	**−0.496***
	Percentage of land covered by shrubs (%)	0.025	0.069	0.263	**0.463***	−0.366
	Percentage of land covered by ground cover (%)	0.391	**0.564****	**0.438***	0.060	−0.381
	Maximum height of trees (m)	0.248	0.158	0.193	0.048	−0.195
	Average height of trees (m)	0.296	**0.702****	**0.727****	0.390	**−0.664****

(continued)

Table 4.11 (continued)

Variable type	Variable name	Individual number	Species richness	Diversity index	Evenness index	Superiority index
	Average height of shrubs (m)	0.383	−0.024	−0.146	−0.218	0.149
	Maximum height of shrubs (m)	0.212	−0.130	−0.210	−0.246	0.257
	Species richness of trees	**0.432***	0.126	0.039	−0.080	0.003
	Species richness of evergreen trees	0.307	0.002	−0.082	−0.086	0.120
	Species richness of deciduous trees	0.410	0.167	0.096	−0.061	−0.061
	Species richness of shrubs	0.068	0.037	0.086	−0.034	−0.034
	Species richness of evergreen shrubs	−0.133	−0.052	0.048	0.028	0.006
	Species richness of deciduous shrubs	0.226	0.064	0.065	−0.221	−0.068
	Species richness of groundcover plants	**0.522***	**0.454***	0.301	0.042	−0.227
	Species richness of plants	0.409	0.226	0.155	−0.039	−0.088

Note *Correlation is significant at a P value of 0.05 (two-sided test). **Correlation is significant at a P value of 0.01 (two-sided test)

This study selects 7 roadside green spaces for in-depth analysis. These roadside green spaces are chosen because they: ① are big enough (with an area above 1000 m^2) to form a certain habitat community; ② border with residential areas to facilitate the

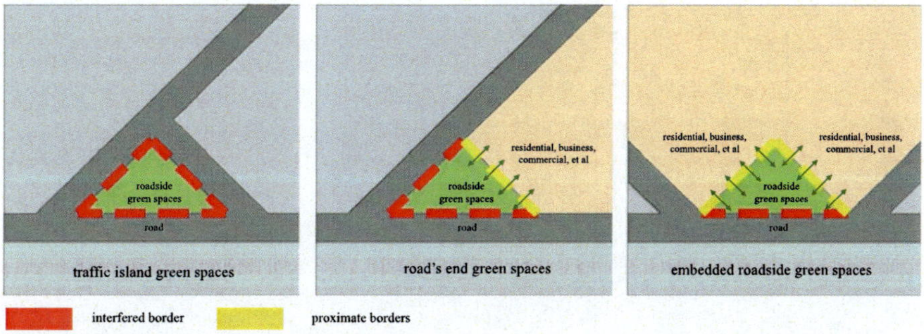

Fig. 4.51 Classification of roadside green spaces along the Century Avenue and analysis of borders

study of the strengthening or weakening effects of these green spaces on the avian diversity in proximate residential areas. Since the selected roadside green spaces are all bordered by land for Class II residential clusters with similar development intensity, the different development intensities of proximate residential areas exert no influence on them; ③ have at least one border that connects with public walkways or the roadways, making it easier to monitor the impact of human interference activities. The locations of the selected green spaces are shown in Fig. 4.52.

4.8.2 *Factors that Interfere with Species Richness of Roadside Green Spaces*

The analysis results of the bird species richness of 7 roadside green spaces show that the species richness of blocks A08-5, A08-7, A10-1, A10-3 with a larger total class area of green space is about half less than that of blocks A13-2, A14-1, and B13-1, and is almost the same as the species richness of proximate residential areas. The species richness of the latter three roadside green spaces is significantly higher than that of proximate residential areas, and therefore has a greater contribution to the avian diversity in the entire block, indicating that the area effect advantage is not obvious at this scale (Fig. 4.53). The study on vegetation structure reveals that the species richness of the 7 roadside green spaces has no obvious correlation with the vegetation coverage, height and type there.

After close observation, we find that the interfered borders of roadside green spaces and the spatial distribution of vegetation may have an impact on species richness. Based on AutoCAD topographic maps and site surveys, we measured the interfered borders and proximate borders of roadside green spaces, and calculated the ratio between the two. When the ratio is greater than 1, interference influence prevails, while when the ratio is less than 1, proximity influence prevails. The results of the study show that the median and mean values of species richness of roadside

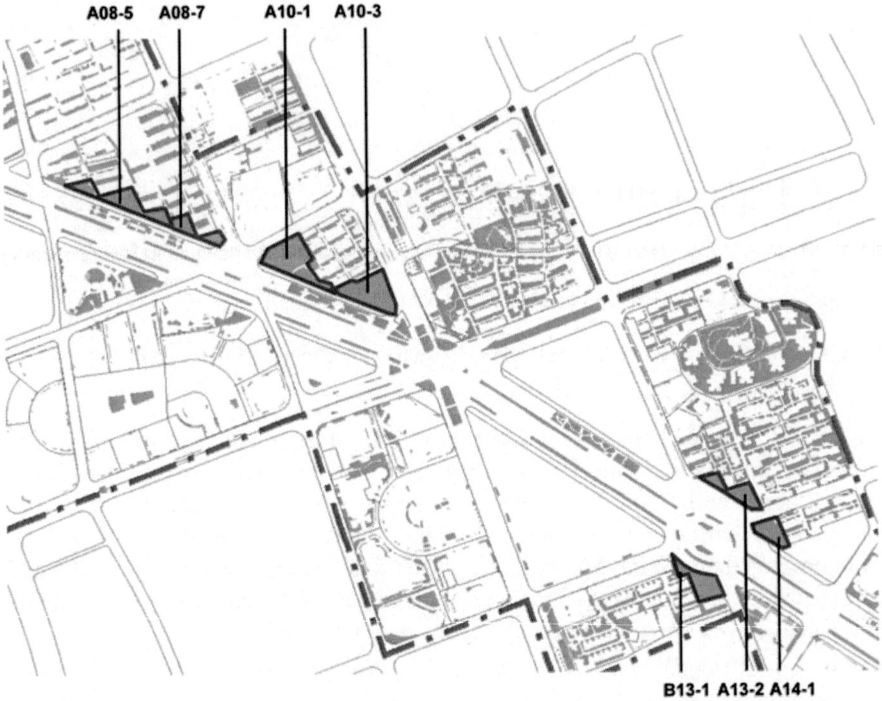

Fig. 4.52 Locations of selected sample blocks in roadside green spaces along the Century Avenue

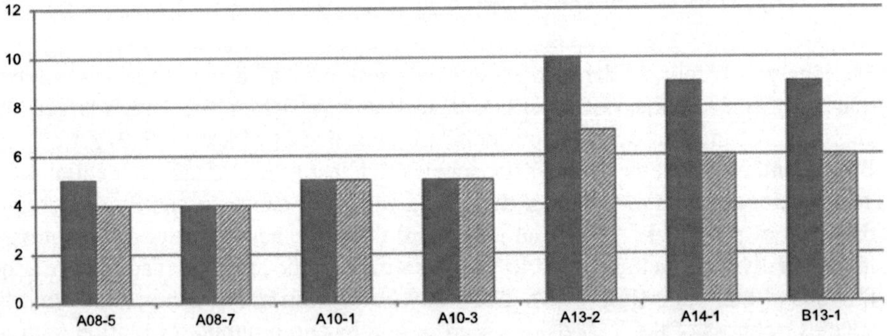

Fig. 4.53 Comparison of the species richness of roadside green spaces and of proximate residential areas along the Century Avenue

green spaces with the value of interfered/proximate borders ratio being less than 1 are both 7, which are greater than the species richness of roadside green spaces with the value of interfered/proximate borders ratio being greater than 1 (median value is 5, and mean value is 6.33). It indicates that shortening the interference borders and extending proximate borders as much as possible will help improve the avian diversity in "stepping stone" street green space patches.

4.8.3 Influence of Spatial Distribution of Vegetation on the Species Richness of Roadside Green Spaces

Among the above 7 blocks, A08-5, A08-7 and A13-2, B13-1 are basically equivalent in the ratio of interfered/proximate borders, but they differ by 50% in terms of species richness. And the bird species richness of the latter two is higher than that of proximate residential areas, while the bird species richness of the former two is basically the same as that of proximate residential areas. Further analysis reveals that this may be related to the spatial distribution of vegetation in the two types of green spaces. The spatial distribution of vegetation in the latter two has a higher canopy density, and the planting position of the first-layer tall trees is closer to the interfered borders (D_t/D in Fig. 4.54 is smaller). Therefore, the tall tree layer has a certain degree of resistance to human interference, and provides abundant and safe habitat space for birds with dense branches and leaves.

4.9 Summary

This chapter takes the avian diversity in the Century Avenue in Pudong New Area, Shanghai as an example, and analyzes in detail the impact mechanism of the urban built environment on biodiversity on the meso-micro scale. The summary is as follows:

(1) The bird community structure in the central area of a high-density city is simple, with resident birds being the majority. Specifically, the species richness of semi-artificial habitats in the attached green spaces is lower than that of semi-natural habitats in the urban parks and public green spaces, while the species richness of residential land is higher than that of land for public facilities. High-rise residential areas with larger centralized central green spaces can have higher bird species richness than multi-storey residential areas with lower development intensity, while high-rise residential areas with only green spaces between houses have the lowest bird species richness. In business areas and commercial areas, the central green space layout is more beneficial to bird species richness than surrounding green spaces and entrance green spaces.

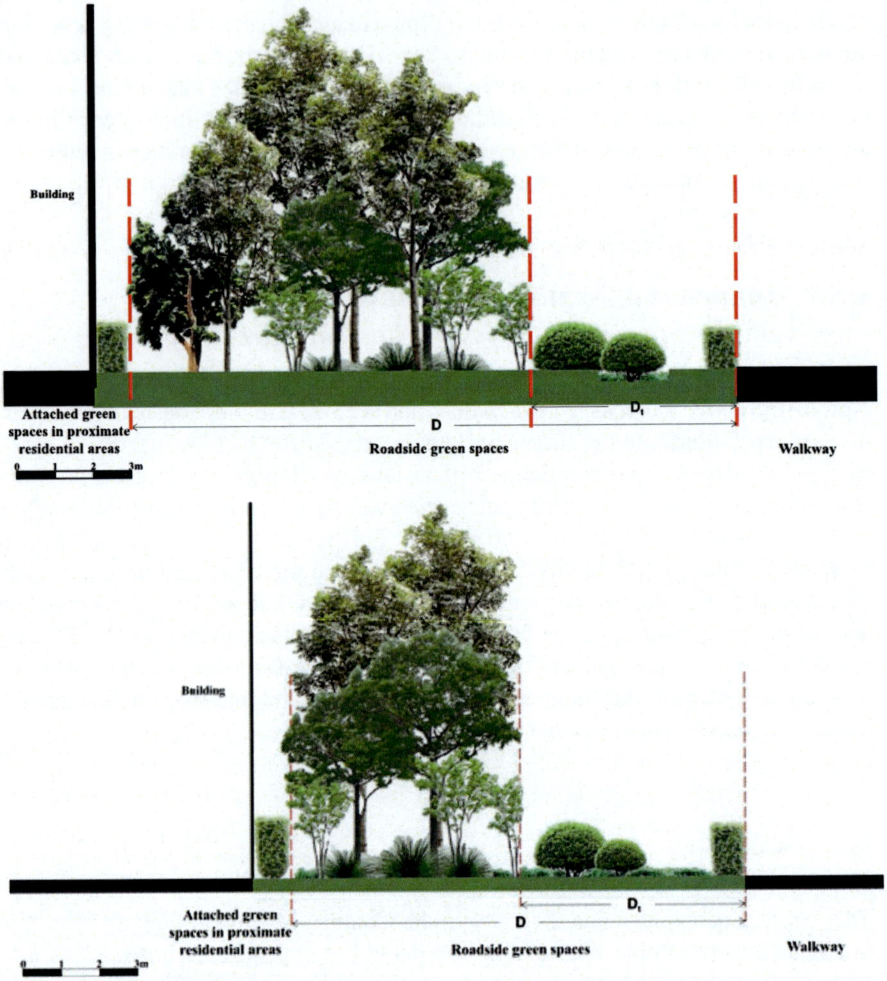

Fig. 4.54 Comparison of the spatial distribution of vegetation in blocks B13-1 and A08-7 (top: B13-1; bottom: A08-7)

(2) Habitats with natural-succession on the abandoned land have higher bird species richness, superior diversity, and more complex community structure of bird species than the surrounding semi-natural habitats in the urban parks and public green spaces and semi-artificial habitats in the attached green spaces, and can become major "sink" habitats or even "source" habitats for urban wild birds. Since such habitats are usually located in land reserves for urban development, they may be destroyed due to a new round of urban development at any time and need special attention in urban renewal.

(3) The nest and feeding niches of bird communities are mainly located in the tree layer and the groundcover layer. The actual microhabitat utilization of birds shows that these two layers are also the most used micro-spaces by birds. In addition, a large number of resident birds of both dominant species and common species stay, rest, and even forage and nest on buildings/structures. In multi-storey buildings, birds tend to stay on the 5th–6th floors and the roof, while in small high-rise and high-rise buildings, they tend to stay on the 12th floor.

(4) Different birds present different preference characteristics in terms of space selection. We need to take into account the proximity effect of birds' feeding and nesting niches in spatial planning and design.

(5) Among the built environment variables, the variables that have a negative impact on avian diversity include: FAR, building density, patch density, perimeter area ratio of green space, largest perimeter area ratio of green space, etc. Among them, building density has the greatest negative impact on avian diversity, whereas FAR has a general negative impact on avian diversity. So, in the development of high-density urban areas, reducing the proportion of building land and providing more surface space for wildlife habitats may be the most effective way to increase their diversity.

(6) Variables that have a positive impact on avian diversity include: green space ratio, water surface ratio, total class area of green space, largest patch area of green space, mean proximity index of green space, patch aggregation index of green space, patch cohesion index of green space, vegetation (trees, ground cover) coverage area, species richness of plants, average height of trees, etc. Among them, the average height of trees has a greater impact, which is a testament to the adaptability of birds in high-density urban areas to off-ground habitats.

(7) In different functional areas, the influence of the shrub layer variables of blocks for urban parks and public green spaces is equivalent to that of the tree layer and ground cover layer variables. The effect of water morphology is more obvious in residential areas and blocks for public facilities.

(8) The avian diversity in roadside green spaces, which are transit "stepping stones" for urban birds, depends to a large extent on the proportion of interfered borders and the retreat rate of first-layer trees. Therefore, it is necessary to reduce the interference of human activities as much as possible and provide a safe shelter space for birds with high canopy density in design.

Urban built environment variables that affect biodiversity on the meso-micro scale and their effects (see Table 4.12 for details).

Table 4.12 Urban built environment variables that affect biodiversity on the meso-micro scale and their effects

Variable dimension		Variable type	Variable name	Influential effect
Biological substrate carrying factors	Ecological land	Land use scale	Total class area of green space (+)	Area effect
			Patch density (−)	
			Largest patch area of green space (+)	
			Patch average area of water (Land for public facilities) (+)	
			Largest water patch area (Residential land) (+)	
		Spatial morphology	Perimeter area ratio of green space (−)	Edge effect
			Largest perimeter area ratio of green space−	
			Mean proximity index of green space (+)	Distance effect
			Patch aggregation index of green space (+)	Network effect
			Patch cohesion index of green space (+)	
	Vegetation	Vegetation area	Tree coverage area (+)	Area effect
			Ground cover coverage area (+)	
		Vegetation structure	Vegetation (tree, ground cover, shrub) species (+)	Height effect
		Vegetation form	Average height of trees (+)	Height effect
			Interfered border/Proximate border (−)	Edge effect

(continued)

Table 4.12 (continued)

Variable dimension		Variable type	Variable name	Influential effect
			Forest border retreat rate (−)	Edge effect
Artificial environment interference factors	Development intensity	Development intensity	Building density (−)	Ratio effect
			FAR (−)	
			Green space ratio (+)	
			Water surface ratio (+)	

Note "+" for positive effect, and "−" for negative effect

References

Chen SH, Ding P, Zhen GM et al (2000) Reserch perspective ecology of urban avian community. Zoo Res 2:165–169 (in Chinese)

Ge ZM, Wang TH, Shi WY et al (2005) Impacts of environmental factors on the structure characteristics of avian community in Shanghai Woodlots in Spring. Zoo Res 26(1):17–24 (in Chinese)

Jiang AW, Zhou F, Qin Y et al (2012) 10-years of bird habitat selection studies in mainland China: a review. Acta Ecol Sinica 32(18):5918–5923

Jin MY (2016) Two years field expedition to find out the "real situation", 80% of 219 bird species in Shanghai are the "passing passengers". Xinmin Evening News, 04-10 (in Chinese)

Lu YW, Tang SX, Shi HL et al (2007) Relationship between Avian community and habitat in Shanghai urban woodlots in winter. Chin J Zoo 42(5):125–130 (in Chinese)

Luan XF (2003) Studies on Avian Community of Shanghai and Planning of Conservation. PhD thesis of East China Normal University (in Chinese)

Ministry of Environmental Protection (2014) Technical guidelines for biodiversity monitoring—birds (HJ 710.4-2014). China Environmental Science Press, Beijing

Shanghai Forestry Bureau (2016) Shanghai completed the second national survey of land wildlife resources. http://www.shanghai.gov.cn/nw2/nw2314/nw2315/nw4411/u21aw1097711.html. Accessed 1 Sept 2016 (in Chinese)

Tang SM, Tang LJ, Li HM (2003) Effects of urbanization on avian community at Wujiaochang, Shanghai. Shanghai Environ Sci 6:406–410 (in Chinese)

Wang YP, Chen SH, Ding P (2000) Effects of urbanization on the winter bird foraging guilds. J Zhejiang Univ (Science Edition) 3: 330–336 + 348 (in Chinese)

Yang G, Wang Y, Xu J et al (2015) The influence of habitat types on bird community in urban parks. Acta Ecol Sinica 35(12):1–13 (in Chinese)

Yuan X, Pei EL, Yan JJ et al (2011) Bird communities and their seasonal dynamics in parks in Shanghai City. J Fudan Univ (Natural Science) 3:344–351 (in Chinese)

Zhang ZT, Zhang ED (2015) Urban wildlife habitat protection model: a case study of Shanghai. Chin J Wildlife 36(4):447–452 (in Chinese)

Zheng GM (2017) A checklist on the classification and distribution of the birds of China, 3rd edn. China Science Publishing & Media Ltd, Beijing (in Chinese)

Zheng WQ (2011) Shanghai Bird Watching Guide. Shanghai wildlife conservation management station, Shanghai Wildlife Conservation Association, World Wildlife Fund (in Chinese)

Chapter 5
Optimization of Urban Planning and Design to Support Biodiversity in Multiple Habitats

5.1 Factors in Built Environment Affecting Urban Biodiversity

Based on the research results in Chaps. 3 and 4, referring to the six effects and according to different spatial dimensions and planning control levels (master plan, regulatory plan, and urban design), the affecting factors in the urban built environment on the macro and the meso-micro scale are divided into the following six categories (as shown in Table 5.1).

5.1.1 The Affecting Factors of Ratio Effect

The affecting factor of ratio effect is mainly embodied in the competition between humans and organisms for the limited space in cities. The negative affecting factors reflecting the occupation of habitat space by urban development activities on the master plan level include: building density, population density, GDP/per km^2, road network density, and ratio of high-rise buildings. The positive affecting factors reflecting the protection of biological space include area proportion of cultivated land, waters and wetland, forest land and major wildlife habitat. The negative affecting factors of urban development activities on the regulatory plan level include FAR and building density, while the ratios of green space and water surface, which have positive effects, indicate that a certain proportion of urban development land provides beneficial habitat quality for biodiversity.

J. Gan, *Urban Biodiversity and Built Environment*,
https://doi.org/10.1007/978-981-16-0949-7_5

Table 5.1 Affecting factors showing different effects in urban built environment on biodiversity on the macro scale and on the meso-micro scale

Scale	Macro scale	Meso & micro scale	
Planning level	Master plan	Regulatory plan	Urban design
Ratio effect	GDP/per km^2 of land (−),building density (−), population density (−), road network density (−), high-rise building area ratio (−), area proportion of cultivated land/water area and wetland/forest/major wildlife habitat (+)	FAR (−), building density (−), ratio of green space (+), ratio of water surface (+)	
Area effect	Area of ecological land (+), area of water area and wetland (+), area of cultivated land (+), area of parks (+) (including area of a single park (+)), area of major wildlife habitats (+)	Area of green Space (+), patch density (−), largest patch area of green space (+), largest water patch area (+)	Coverage area of trees (+), Coverage area of groundcover plants (+)
Edge effect			Perimeter area ratio of green space (m/m^2) (−), perimeter area ratio of the largest patch of green space (−), interfered border/proximate border (−), Setback rate of woods' edge (−)
Distance effect		Waterfront-forest proximity (+)	Mean proximity index of green space (+)
Network effect	Water network (+)	Patch aggregation of green space (+)	Patch aggregation index of green space (+)
Height effect			Species richness of vegetation (+), Average height of trees (+)

Note + for positive affecting factor, and − for negative affecting factor

5.1.2 The Affecting Factors of Area Effect

The affecting factor of area effect is mainly reflected in the minimum area for biological habitation and population succession, so it basically generates positive effects except for patch density. The affecting factors on the master plan level mainly lie in guaranteeing a certain area of ecological land to maintain the carrying of organisms and the ecological security pattern, including: the area of ecological land and the area of water and wetland in it, the area of cultivated land, the area of parks, the area of major wildlife habitats, as well as the area of a single park. The area effect continues on the levels of regulatory plan and urban design. The affecting factors on the regulatory plan level include: the total area of green space patches, the largest patch area of green space, the average patch area of water, and the largest water patch area; and the affecting factors on the urban design level mainly include coverage area of trees and groundcover plants. Therefore, from the use of green space on the macro scale, to the property of green space on the meso scale, and then to the coverage of vegetation on the micro scale, the area effect on the whole reflects the scale demand of habitat space.

5.1.3 The Affecting Factors of Edge Effect

In the urban environment, the edge effect factors mainly affect biodiversity by reducing the interference of human activities to biological space, and therefore are all negative. As the edge effect is mainly determined by the form of green space, it mainly works through urban design on the meso-micro scale. For green spaces within a block, the affecting factors include the perimeter area ratio of green space (m/m^2) and the perimeter area ratio of the largest patch of green space (m/m^2); for green spaces around the street corner or roadside green spaces, the affecting factors include the interfered border/proximate border ratio and the setback rate of woods' edge.

5.1.4 The Affecting Factors of Distance Effect

The distance effect means that the degree of proximity affecting the movement and material and energy exchange of organisms between different patches. The main affecting factor includes the average proximity of green space patches within a block on the urban design level. In addition, according to the research in Sect. 4.4.4, the proximity of water banks and forest clusters in the waterfront zone, as a spatial factor supporting aquatic birds' preference for proximity of feeding and nesting niches, can also be included on the regulatory plan level. In fact, numerous studies have proved the impact of landscape proximity of various ecological lands on biodiversity on the macro scale. In this study, we find it difficult to regulate the proximity of various ecological lands on the master plan level in the built-up areas, and therefore exclude the distance effect on the master plan level from the analysis framework.

5.1.5 The Affecting Factors of Network Effect

The network effect means systematically linked corridor networks affect the migration and daily activities of organisms between patches and non-inbreeding across patches. That is especially important for the urban fragmented habitats which have small patches and are far away from each other. The affecting factors on the master plan level exert influence mainly through the relatively natural water network landscapes, while those on the regulatory plan and urban design levels through the aggregation and combination of green landscapes.

5.1.6 The Affecting Factors of Height Effect

Height effect is a special effect of the biological spaces in high-density cities. Research in Chap. 4 shows that biodiversity on a high-rise and low-density block is not necessarily worse than that on a low-rise and high-density block. When space resources are limited, the building surfaces created by three-dimensional spaces can also provide new off-ground habitats for organisms. Taking birds as an example. Due to their physiological need of flight, the demand of birds for vertical niches is misaligned with human activities. Studies have confirmed that birds make use of the vertical surfaces of buildings, so three-dimensional greening and their locations (on which floors) can be factors representing the height effect on the urban design level. Furthermore, the height and species of vegetation (especially of trees) are also important factors due to birds' demand for the feeding, nesting, and resting niches on the tree layer.

5.2 Planning and Design Regulation Model Based on Biodiversity Performance in Urban Built Environment

5.2.1 Comprehensive Assessment of Biodiversity Performance in Urban Built Environment

Performance means effect, result and achievement, and also implies the meaning of efficiency. Yan et al. (2012) define "the environmental performance of urban spatial structure" as "the environmental effect or performance generated by the spatial distribution of urban elements and their interactions, including energy flow, material cycle, and biodiversity." Based on that, this study defines "the biodiversity performance in urban built environment" as **"the effect of spatial distribution of urban built environment elements and their interactions on biodiversity.**"

According to the theoretical analysis of the relationship between urban biodiversity and built environment in Chap. 2 and the empirical researches in Chaps. 3 and 4, the affecting factors in built environment on urban biodiversity include the habitat quality factors and human-induced disturbance factors. This study therefore defines a comprehensive assessment model for the biodiversity performance in the urban built environment, namely the Urban Biodiversity Performance (UBP), Assessment which assesses the biodiversity performance in the built environment within a certain urban space from the perspectives of Q (Habitat **Q**uality) of and D (Human-induced **D**isturbance).

As shown in Fig. 5.1, the vertical axis represents the value of Q, which is the assessment index for the spatial quality improvement of the habitat within a certain boundary, and the horizontal axis represents the value of D, which is the assessment index for the disturbance by human activities within and around a certain boundary.

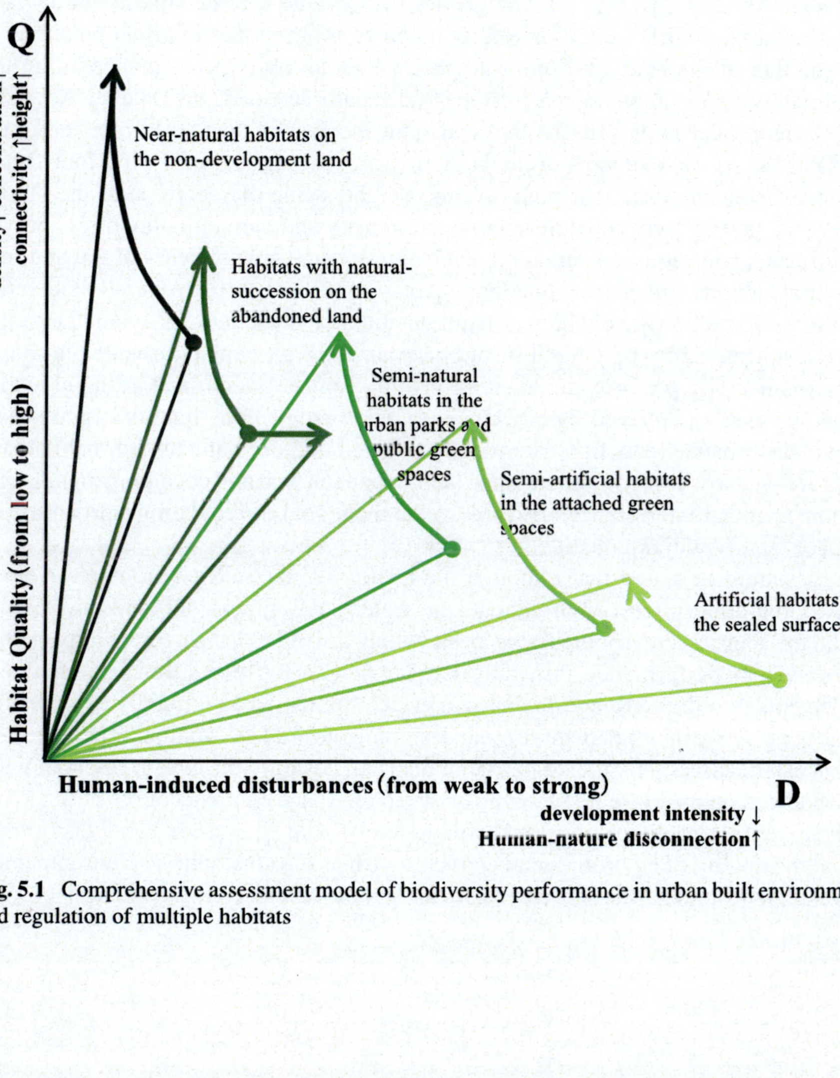

Fig. 5.1 Comprehensive assessment model of biodiversity performance in urban built environment and regulation of multiple habitats

Urban Biodiversity Performance (UBP) is represented by the slope of a straight line through the origin. The higher the slope (i.e., the steeper the straight line), the better the UBP.

5.2.2 Planning and Design Regulation Objectives to Improve Biodiversity Performance in Urban Built Environment

In the comprehensive assessment model of the UBP, the near-natural habitats on the non-development land, the semi-natural habitats in the urban parks and public green spaces, the semi-artificial habitats in the attached green spaces, the artificial habitats on the sealed surfaces, and the habitats with natural-succession on the abandoned land have different disturbance and quality ranges, thus presenting different slope grades. As shown in Fig. 5.1, the greater the Q value and the smaller the D value, the greater the UBP value. Therefore, it can be inferred that in urban planning, the regulation of the built environment space for biodiversity performance in multiple habitats is mainly about increasing the Q value and decreasing the D value. According to the empirical studies in Chaps. 3 and 4, the increase of the Q value depends on the rise of the spatial elements of the built environment and the following four factors: density, concentration, connectivity, and height; while the decrease of the D value depends on the decrease of development intensity and human-nature disconnection. Moreover, conscious conservation and increase in the area of near-natural and semi-natural habitats with greater overlapping may contribute more to overall biodiversity. Empirical studies prove that near-natural habitats with better quality and larger area are less affected by peripheral disturbance. In the high-density urban environment, the biodiversity performance of semi-natural, semi-artificial, and artificial habitats that are easily perceived by citizens has a wide range, and "quality improvement" and "disturbance reduction" are mainly targeted at those habitats. For habitats with natural-succession on the abandoned land, urban renewal and development may bring more artificial interference pressure, so the main goal of regulation is to ensure that space quality will not decline.

It should be noted that although the quality of the built environment elements (e.g., the atmospheric environment, the water environment, the soil environment, and the acoustic environment) has been widely proved to have a direct impact on the biodiversity performance, this study does not include those traditional environmental elements into the UBP comprehensive evaluation model as it mainly explores from the perspective of urban planning and designing the built environment space elements that can be directly regulated in spatial plan and design. Subsequent research will be conducted on the close relationship between the space elements of the built environment on the two dimensions of "habitat quality" and "human-induced disturbance" and the quality of environmental elements such as the atmospheric environment, the water environment, the soil environment, and the acoustic environment.

5.3 Principles of Optimization in Planning and Design to Enhance Biodiversity in Urban Built Environment

Urban planning aims to make reasonable allocation and overall arrangement of urban space resources for the future development of cities, and needs to consider the development needs on ecological, social, and economic dimensions. As living things inhabiting in cities other than humans, organisms should enjoy the urban space resources just like humans do. Introducing biodiversity into the urban spatial planning system, and including the protection and enhancement of biodiversity in the urban built environment into urban planning are not only conducive to the survival and reproduction of organisms, but also can create a more livable environment for humans, and make the complex urban ecosystem more stable and resilient to external disturbances. According to the current planning levels and the planning objectives and contents of each level, as well as the six effects on biodiversity, basic optimization principles are put forward on the master plan, regulatory plan, and urban design levels to improve the biodiversity performance in the urban built environment (Table 5.2).

5.3.1 On the Level of Master Plan: Quantity Assurance, Zoning, Level Classification, and Networking

Urban master plan aims to coordinate the overall allocation and structural arrangement of urban space resources at the macro strategy level of a city. Against the backdrop of high-density urban development in China, when it comes to improving biodiversity performance in the urban built environment on the master plan level,

Table 5.2 Principles of optimization in planning and design to improve biodiversity performance in urban built environment and their effects

Effect	Master plan	Regulatory plan	Urban design
Ratio effect	Quantity assurance	Size increase	–
Area effect	Zoning	Greening centralization	–
Edge effect	Level classification	–	Disturbance reduction
Distance effect	–	Distance control	–
Network effect	Networking	Corridor access	Benefit improvement
Height effect	–	–	Planting suitability, symbiosis

the key lies in "protection." Specifically, facing such urban development constraints as land, water, and the environment, efforts should be made to maintain and protect the residual natural and near-natural habitats in the limited urban resources, which have not yet been eroded by the artificial environment, create a spatial structure which can help achieve the dynamic equilibrium between the construction of urban human settlement and the inhabitation of organisms, so as to maintain the stability and evolution of the entire ecosystem. It involves the four principles below.

(1) **Quantity assurance**: The process of urban development by mankind is to some extent a process of encroaching on the homeland of the protozoa. Therefore, the top priority of urban planning and design, which aims to support biodiversity, is to reasonably distribute the "ownership" of land space resources for human activities and biotic inhabitation. Priority should be given to the habitat space, which should be regarded as one of the bottom lines of planning resources allocation. Targets of basic space thresholds should be set to guarantee the existing biological resources so that among all types of land, the land resources whose major ecosystem function is to conserve biodiversity account for a significant proportion. Balance should be achieved among various types of ecological land, including waters, wetland, cultivated land, and woodland, ensuring no further encroachment caused by human development activities in long planning terms. Additionally, relevant requirements will be included in the indicators of urban growth boundary and development land size.

(2) **Zoning**: Based on the identified demand for biotic space resources, adequate information about biodiversity, and the benchmark of biotic spaces in urban-rural land, ecological function areas should be zoned to identify the major natural and near-natural biological habitat patches which are irreplaceable for biodiversity protection, determine their location, scope, and size, form the peripheral ecological environment barrier, and incorporate them into the ecological protection red line in the prohibited, restricted, and permitted areas for urban development.

(3) **Level classification**: The protected areas should be divided into different protection levels, and space control measures (for prohibited and restricted construction) and restriction of the frequency of human activities on each level should be identified. Requirements should be put in place for safe distances between the protected areas and surrounding communities, and for approved development projects. For the semi-natural habitats outside the protected areas, such as large green spaces in urban parks, it is suggested to make regulatory requirements for approved development from the dual perspective of interaction between human recreational activities and biotic inhabitation, and consider them as supplementary biotic habitats, which, together with the protective habitats, can form a protection system.

(4) **Networking**: Isolated habitat areas can only maintain the survival of a single species, but prevent gene exchange and energy exchange between populations. Therefore, the ecological principle of aggregate-with-outliers can be applied to connect the "stepping stones" along the main natural boundaries of natural, near-nature, and semi-natural patches on different levels and of different sizes, thus forming networked biotic spatial substrates. Human activity zones can be arranged in the "enclaves" along the main natural boundaries, to reduce interference with the major natural habitat areas and guarantee the survival, migration, and evolution of biological species.

As the first stage of urban planning, biodiversity improvement planning on the master plan level paves the way for planning, construction, and management at the next level.

5.3.2 On the Level of Regulatory Plan: Size Increase, Greening Centralization, Distance Control, and Corridor Access

Regulatory plan aims to regulate the nature, intensity of use, and space environment of development land. It plays a significant role in regulating the ecological environment of a city by conveying the objectives, demarcating the boundaries, and identifying the indicators for regulation. Biodiversity planning at this level is the weakest link of the current planning system, thus requiring the most efforts. According to the basic function of the regulatory plan, that is, defining the bottom line, boundary, and quantity, there are four main objectives for biodiversity planning on this level.

(1) **Size increase**: In light of the development goals identified in the master plan and the regulatory requirements for major protective biotic habitat patches and for supplementary biotic habitat patches, implementing the indicators of different levels in the centralized built-up area and the non-centralized built-up area, the size of the green infrastructure carrying habitats should be expanded as much as possible amid high-density and high-intensity development, and different control indices and construction requirements regarding ecological infrastructure should be made for various function areas.

(2) **Greening centralization**: Advocating the ecological green space system architecture with centralized green infrastructure as the core, specifying the requirements for location, scale, and shape accordingly, and regulating important green space and waters by incorporating them into the urban green line and blue line.

(3) **Distance control**: Identifying the distance between ecological green spaces with different functions of carrying biotic habitats according to the action radius of biological species and the approaching demand for feeding and nesting spatial niches.

(4) **Corridor access**: Connecting green infrastructures of different scales through the linear ecological corridors along roads and rivers, building ecological green networks in the form of points, lines, planes, and corridors, and specifying requirements for the grade, function, width, and construction of corridors.

5.3.3 *On the Level of Urban Design: Benefit Improvement, Planting Suitability, Disturbance Reduction, and Symbiosis*

Urban design is the overall conception and arrangement of the urban space, and is an effective means to implement urban planning, guide architectural design and shape urban form. From the perspective of improving biodiversity performance in the urban built environment, urban design mainly serves to adjust the various morphological patterns directly facing habitat space, provide on the ground and in the vertical space potential feeding, nesting, and resting niches required for the survival of organisms, and form a networked micro-natural system that improves the capacity of human living spaces to bear biological inhabitation. It involves the four principles below.

(1) **Benefit improvement**: Studying the ecological benefits of the micro-habitats within a block based on the leading functions of different ecosystem services, and identifying the construction and renewal requirements for improving the ecological benefits in combination with the visual function of landscapes.
(2) **Planting suitability**: Making requirements for planting in habitat patches, including species of vegetation, structure of layered planting, vegetation height, selection of seeds, etc.
(3) **Disturbance reduction**: Proposing spatial arrangement and vegetation distribution with reduced human disturbance for urban biotic habitat patches highly overlapping with the human settlement, so as to ensure that the biological space and the human settlement are separate yet not divided.
(4) **Symbiosis**: Creating on artificial structure surfaces some micro-habitat spaces for organisms to stay, feed, and even nest, and making requirements for increasing the surfaces of living spaces for organisms, thereby realizing spatial-temporal displacement and win-win symbiosis between organisms and humans in three-dimensional urban spaces. Green roofs, green walls of some height, and green facades can provide habitats and planting environments free from human influence. From the perspective of biodiversity, the underlying benefit is that an off-ground planting environment is less susceptible to intensive human disturbance and has the potential to become a sanctuary for biodiversity. For example, green roofs keep birds away from predators such as cats and dogs that move on the ground.

5.4 Planning and Design Optimization Strategies to Improve Biodiversity Performance of Urban Built Environment

5.4.1 *Master Plan*

(1) **Ascertaining a minimum threshold space of ecological land and define the detailed proportion of each ecological land category**

The non-compact use of land resources for urban development over the past four decades has driven the proportion of ecological land in Shanghai and other metropolitan areas in China close to a bottom-line maintaining urban ecological security. Therefore, during the planning process, determining a threshold minimum space for ecological land should be prioritised to maintain biodiversity, and the construction land space should be configured.

The proportion of ecological land for an international metropolis generally exceeds 50%. A level equal to more than 60% is also suitable (Li et al. 2014). According to the New Master Plan, 'Ecological land accounts for at least 60% of the total land area by 2035' (Shanghai Environmental Protection Bureau et al. 2013). However, a detailed action plan for the realisation of this target is still missing. From the perspective of biodiversity protection, those lands that serve a certain natural and near-natural ecosystem conservation function must be restored, and their proportion must be gradually increased over time.

The proportion of ecological land corresponding to each ecological land category (e.g. forest land, wetland, cultivated land and public green space) can be further detailed in accordance to the ecological land composition of different metropolitan areas and biodiversity-supporting services. For instance, in Shanghai where more than 70% of wild vertebrate species are living in coastal wetlands, the wetland resources preservation rate can be used as a detailed index of ecological land. Meanwhile, for those cities where mountain land serves as the main habitat space, the mountain forest coverage rate must be defined as the detailed index.

(2) **Establishing a multidimensional network for the protection of multiple habitats: "nature reserves-major habitats-supplementary habitats-symbiotic habitats"**

Based on the current management and control of animal wildlife habitats (nature reserves mainly), major wildlife habitats should also be included in the relevant ecological protection management and control, with strict prohibitive and restrictive regulations being imposed on development. In addition to the reserves and major habitats, wilderness parks in the suburbs should be added as supplementary habitats, and urban recreational parks with a larger area and better ecology in the city center should be regarded as symbiotic habitats for humans and organisms. A four-level protection network is therefore established, where biodiversity protection is a shared mission for all. Control of space development and restriction of the frequency of human activities vary for the habitats of different levels. Meanwhile, species in the habitats of the four levels will be continuously monitored and a dynamic database connected to the spatial information base will be established.

Major wildlife habitats are essentially smaller nature reserves, and their purpose is to make the wildlife habitats outside the suburban nature reserves free from the interference of human activities as much as possible, and also facilitate the development of surrounding communities. As the development in China's major cities moves from the city center to the suburbs, the disappearance, segmentation, or fragmentation of suitable habitats for wildlife in the suburbs is inevitable during the process of urbanization (Tang et al. 2003). Therefore, strict demarcation of major habitats for wildlife has been put on the agenda in some cities. Take Shanghai for example. From 2013 to 2015, to meet the requirements of strengthening ecological protection in the report of the 18th National Congress of the CPC, the city incorporated the development of major wildlife habitats and the restoration of small populations into its three-year plan for forestry development.

Unlike general parks which offer people space for recreational activities, the first and foremost purpose of urban wilderness parks is to protect and build habitats for wildlife. Wild parks are also named as country parks, nature parks, or ecological parks. For example, 21 country parks have been preliminarily planned in the suburbs of Shanghai, with an area of 3,500 km^2 of planned ecological land. By the end of 2016, five country parks had opened successively. Wild parks can also be used as supplementary habitats for wildlife. Like major wildlife habitats, wild parks should be large enough to ensure the habitation and reproduction of a certain proportion of biological communities. This approach has been adopted by many scholars. Through their study of Madrid, Osaka, and other cities in America, Slovakia, and Poland, Fernández-Juricic and Jokimaki (2001) find that the minimum park area required by birds in those cities is between 10 and 35 hm^2. This provides biological standards for the design of urban parks which focuses on the establishment of habitats for birds. Planners can set the minimum area of a supplementary habitat according to the requirements for the inhabitation of target species and for the size of the buffer zone and the tourist activity area.

(3) **Improving urban ecological river network connectivity**

Chapter 3 proves the effects of the network closure (α), the line-point ratio (β) and the network connectivity (γ) of water landscape on regional biodiversity.

In the river and lake system, ecological rivers are often believed to serve as habitats, conduits, filters, sources, and sinks for their reasonable ecological system structure and excellent functions and for their tenacity, stability, and resilience against long-term or sudden disturbance. Ecological rivers have an irreplaceable role in the development of urban ecological network, and they can form a complex ecological base together with road and green land, to connect important habitat nodes, and give full play to the network effect. Ecological rivers can either be natural or ecologically restored. The connectivity of ecological river networks shows the network structure of rivers with ecological corridor effect. During the urban ecological network creation, much attention should be paid to: (1) building urban water network nodes with certain scale and ecological corridors with higher connectivity; and (2) the increase of accessibility to ecological landscape patches, in order to meet the demands of habitat, migration and gene exchange.

(4) **Appropriately controlling development intensity and avoid development construction intrusion in key natural land areas**

The research results in Chap. 3 testify to the fact that excessive development density and intensity have negative effects on biodiversity on the macro scale. The development behavior of humans is the factor which has a biggest influence on biodiversity. Mammals and fishes are more sensitive to the environmental change generated by development intensity, when compared to other species. Large-scale development also brings a lot of invasive species. The objective of development control at the macroscale is to designate built and non-built areas. Land resources for non-built areas don't necessarily need to be used for biodiversity conservation, but space has to be reserved at least. Therefore, when considering the protection and enhancement of biodiversity at large scale, we shall firstly define a certain limitation for the proportion of occupied space for human activities, prevent artificial construction from expanding disorderly, and reserve appropriate space for wildlife habitat.

5.4.2 Regulatory Plan

(1) **Creating the green space pattern in semi-artificial habitats based on the "sink-corridor-stepping-stone" habitat chain system**

This study proves that the diversity of birds benefits from concentrated and highly connected green space patches which are close to one another. In semi-artificial habitats in the attached green spaces that cannot provide much large-scale greening land, it is necessary to increase the connectivity of urban landscape by setting up small-scale green space patches to provide "stepping stones" and "transit stations" for the free movement of animals. Take urban residential-commercial areas for example. The author proposes a green space pattern featuring hierarchical concentration and connection. Specifically, "stepping stones" of different levels can be formed by connecting via corridors large concentrated green spaces, medium-sized green space clusters, and small or micro green spaces such as roadside green spaces, roadside green spaces, and subsidiary green space. The largest concentrated green space, as the "sink" habitat, provide feeding space and even nesting space needed by species, while green space clusters, as the rest stop for organisms moving in a block, enhances the communication of species within the block through the corridor. Roadside green spaces, street-side green spaces, and roadside green spaces jointly serve as the transit stations for cross-block migration and gene exchange of species (Fig. 5.2).

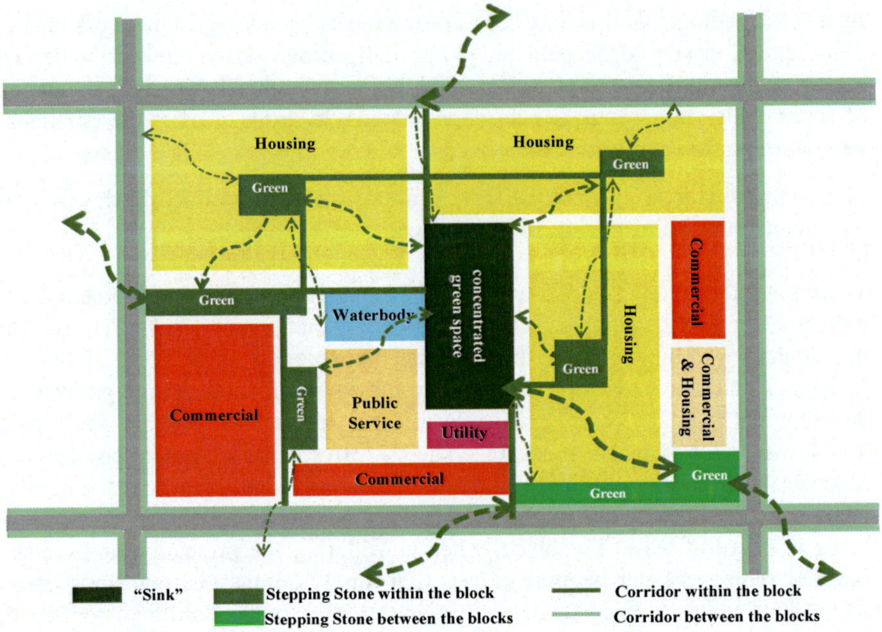

Fig. 5.2 Habitat chain system in the urban residential-commercial area

The largest concentrated green space ("sink") and green corridors ("corridors") are the key elements of the habitat chain system.

(1) The largest concentrated green space

Concentrated green spaces refer to large patches of concentrated green belts, including green blocks or linear green spaces except for green spaces between houses and roadside green spaces. Take residential areas for example. According to the *Planning and Design Specifications for Urban Residential Area (2002)* (GB 50180-2002), the width of public green blocks or linear public green spaces should not be less than 8 m, their area should not be less than 400 m^2, and at least a third of green spaces should lie beyond the standard shadow range on buildings produced by sunlight. The largest concentrated green space refers to the largest green block or belt among the concentrated green spaces within a block, and generally serves as the central green space. The largest concentrated green space with a certain scale is more beneficial to biodiversity and has greater potential to become a "sink" habitat.

(2) Green corridors

In the urban environment, even though the horizontal distance between two biotic habitats seems close, buildings or structures may stand in their way. Therefore, apart from narrowing the distance, efforts must be made to strengthen the role of corridors. The green corridors at the regulatory level mainly include linear spaces such as greenways, streets, walking paths, and river corridors, which serve as the "stepping stones" to connect concentrated green blocks, linear concentrated green spaces, and punctate green spaces. In the planning, the level and the main ecosystem services of road and river corridors can be determined respectively according to the current situation and the number of connecting nodes of corridors, and the width of corridors can be determined based on the migration and movement demands of target species, such as meeting the activity needs of birds and population stabilization of native herbs. Green corridor networks can be put under control by being incorporated into the boundaries of the green and blue lines.

(3) **Encouraging ecological construction on artificial water surfaces**

As an important part of green infrastructure, waters provide necessary spatial niches for urban organisms to drink water. In this study, patches with water surfaces have better biodiversity in general. In addition to attracting birds that depend on the aquatic environment, such as game birds and waders, wood-birds are often found drinking and staying near water. The analysis in Chap. 4 also proves that in residential areas and public facilities, the total patch area, the largest patch area, and the perimeter area ratio of water are significantly correlated with species diversity. However, in the investigation, the author also found that most of the water bodies in urban blocks are stagnant, with no circulation. Without proper management, they will easily dry up and repel both living things and people. The author believes that in cities like Shanghai boasting abundant water resources, developers should be encouraged to build artificial water surfaces with ecological benefits and biological support functions, and imitate natural water habitats in water landscape design, thereby elevating the function of water bodies from pure landscape to natural circulation.

(4) **Afforestation within a certain range of the riverside area**

River banks and intertidal zones provide feeding space for waterfowl such as Black-crowned Night Heron. In the ecological construction of the riverside area, to narrow the distance between the feeding space and nesting space for this particular species, a certain size of woods should be retained or planted in proper locations near the banks. That should be included into the regulatory plan as one of the requirements for green space distribution in the waterfront area. The distance of woodlands from the banks should be less than the foraging range of target species.

(5) **Highlight the value of the natural capital in urban renewal on abandoned land after natural succession**

This study finds that the abandoned natural-succession land tends to have better biodiversity than surrounding areas. However, as urban development proceeds, some abandoned land reserved for development will be put into use again. Take the small wetland beside Shanghai Science and Technology Museum mentioned in Chap. 4 for example. As an area reserved for the cultural facilities of the Huamu Administrative and Cultural Center, the wetland has become the site of the East Wing of the Shanghai

Museum. There is nothing wrong about the planning process, but as the only open wetland in central Shanghai for the reproduction of more than two kinds of wild water birds, the small wetland cannot be replaced by semi-natural parks or semi-artificial recreational green space in terms of its biological support value. The international design competition for the East Wing has now completed and the winning design also seeks to preserve the existing ecosystem to the maximum extent possible, but it is unlikely to succeed given the tremendous scale of the development project. Therefore, the author believes that, in the urban renewal, the value of the land that has been abandoned for more than a decade and may have gone through the natural succession and the natural assets on it should be assessed before determining the project to be introduced and the development scale, so as to strike a balance between development and protection, since this special kind of urban habitat makes significant contributions to the urban biodiversity.

5.4.3 Urban Design

(1) **Optimizing the shape and spacing of centralized green spaces**

Compared with linear green spaces between houses and on roadsides and punctuate street green spaces, concentrated green spaces perform better in supporting the natural ecosystem, and are easier to form relatively reclusive green environments, hence becoming the smallest inhabitation units for organisms. The shape and spacing of concentrated green spaces are crucial for the biological substrate environment's quality. The results in Chap. 4 show that the smaller the marginal/area ratio of the largest concentrated green space is, the more favorable it is for avian diversity. Therefore, under the prerequisite of guaranteeing the size of the largest concentrated green space, efforts can be made to further optimize its shape: a circular green space is better than a square one, and a square one is better than a rectangular one. With the same area, the smaller the landscape metrics index (perimeter area ratio) is, the less susceptible it is to external artificial interference.

In the past, green space distribution, "green space service radius" or "green space accessibility" were generally adopted as distance indicators, with more consideration given to human activity distance than to biotic activity range. In order to improve the biodiversity performance of the urban built environment, it is necessary to maintain the distance between concentrated green spaces inhabited by organisms within the radius of biotic action competence, reducing the obstacles of biotic movement and increasing exchange and reproduction of species. Figure 5.3 shows the action distance of some species. For example, beetles, ants, and shrikes will not go beyond 50 meters away from hidden green spaces, and grubs and frogs will not go beyond 150 m away.

Fig. 5.3 Distance of movement of animals away from hidden green spaces. *Photo source* Redrawn from Lin (2001)

(2) Promoting multi-layered tree-shrub-herb planting structure and food-borne native vegetation

Research in Chap. 4 shows that multi-layered plant distribution is more conducive to wild birds. Multi-layered vegetation constitutes a diversified living environment, where the multi-layered community pattern, dominated by evergreen trees and combined with deciduous trees, shrubs, and groundcover plants, allows birds to live in the canopy, build nests in the branches, and walk among shrubs. Different kinds of birds and insects move at different heights and locations, and the multi-layered planting approach enables each vegetation layer accommodate numerous organisms. As the tree layer and the ground layer constitute the major nesting and feeding spatial niches for organisms, more attention should be paid to the distribution of these two layers.

When it comes to plant distribution, the number of species of trees should be determined first. Relevant studies show that, for bases above 2 hm^2, the minimum ideal plant diversity can be achieved with more than 20 kinds of the lowest trees and more than 15 kinds of the lowest shrubs (Lin 2007). In addition, the tree-shrub-herb ratio in nature is 1:7:22, showing a high proportion of groundcover plants, yet in the current urban green space system planning, when determining the tree-shrub-herb ratio, the proportion of tree species is often set too high and that of herbaceous plant species too low (Hao and Zhang 2015), which has seriously restricted the improvement of plant diversity in urban green spaces and also affected the diversity of other species. Following the laws of nature, we should increase the number of shrub and herb species as much as possible while guaranteeing the number of tree species, so as to ensure a tree-shrub-herb ratio close to that in nature. Natural or near-natural approaches should be adopted in planting design to imitate natural ecological communities and follow the laws of nature, thereby improving the naturalness of vegetation landscape.

Given that plants provide food niches for birds and other species, plant distribution in large green spaces, green corridors, and surrounding three-dimensional greening should make use of the rivalry between living things on the biological chain, attract

Table 5.3 Foodborne native plants in the middle and lower reaches of the Yangtze River and surrounding areas

Type of plant	Foodborne plants
Trees	Chinese Hackberry, Paper Mulberry, Mulberry, Chinese Bishopwood, Beadtree, Chinese Tallow, Glossy Privet, Heather, Winterberry euonymus et al.
Shrubs	Wild Cherry, Firethorn, Tricuspid Cudrania, Pittosporum, Border privet
Climbing or sprawling shrubs	Wild Rose, Wintercreeper, Honeysuckle, Ivy
Groundcover plants	Broadleaf Liriope, Indian Strawberry

Source Tong et al. (2009)

more diverse species of organisms with food-borne native plants, and provide more food spatial niches. See Table 5.3 for the foodborne native plants available in the middle and lower reaches of the Yangtze River.

(3) **Using layered and three-dimensional greening as off-ground habitat patches and corridors**

This study has preliminarily demonstrated the potential of three-dimensional surface substrates to provide habitats for urban wild birds. Due to the increasing shortage of land in high-density cities, the adoption of new technologies such as three-dimensional greening to increase the green surface area of three-dimensional spaces has become an important measure for urban ecological construction in many cities. However, most researches and practices focus on the visual effects and maintenance of three-dimensional greening, paying little attention to its ecological performance, especially the biodiversity performance. Some foreign studies show that off-ground habitats created by roof gardens can provide space for the pollination, foraging, shelter, and protection of wildlife, and original resources, which has a wide positive effect on biodiversity and the ecological environment (Brenneisen 2006); vertical green walls can also serve as important potential urban habitats to support the survival of rich species (Francis 2011).

More green surfaces, which provide off-ground micro-habitats for organisms, will be available by encouraging three-dimensional greening through financial subsidy, conversion of green space rate, and incentive of FAR. Based on the functions of buildings and the height of surrounding trees, vertical "stepping stone" habitats can be created by increasing the likelihood of birds and other species to forage and nest at an appropriate height of three-dimensional greening. For example, by setting up hanging gardens of different levels in the design of a residential building, hanging foodborne vegetation grooves in the building facade, and climbing plants or green walls and other migration corridors between the roof and ground green space, and connecting them with surrounding ecological habitats, a three-dimensional green conduit will be available for animals such as insects and small lizards, improving the biological efficiency of multi-layered greening (Figs. 5.4 and 5.5).

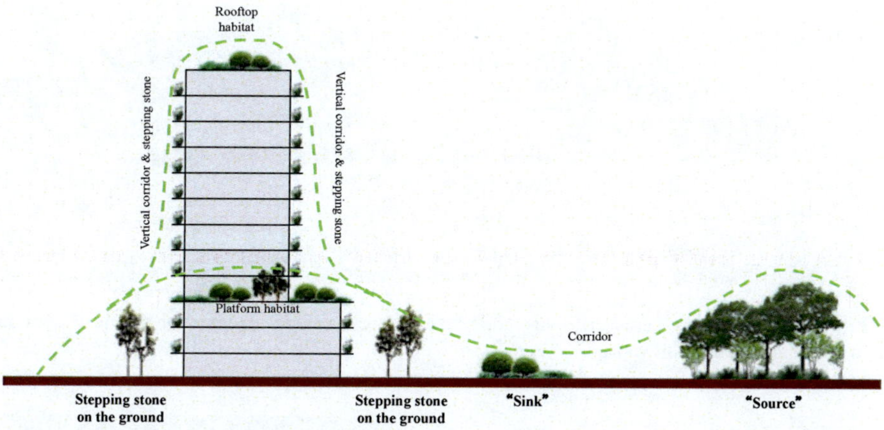

Fig. 5.4 Off-ground habitat patch and corridor system

Fig. 5.5 Examples of off-ground habitat patch and corridor (Left: Bird house at the Botanical Roof Garden of Augustenborg, Malmo; Right: The vertical "stepping stone" habitat in a residential community of India)

(4) **Considering the preference of birds to nest in holes in the design of urban buildings, structures, and street furniture**

Some urban wild birds have the habit of using building holes for nesting. Therefore, in the design of urban building facades, sculptures, and street furniture, holes can be consciously reserved to provide nesting space for relevant birds. For example, in new and renovated buildings, Germany consciously provides nesting space for swifts, pigeons, and bats that depend on buildings to inhabit (Fig. 5.6) (Werner et al. 2016). Another example is a thatched sculpture in Jing'an Sculpture Park, Shanghai, which attracts a large number of sparrows to build their nests (Fig. 5.7, left). Similarly, Beijing Olympic Park once built a swift tower, the first artificial structure built for animals in China, with a total of 2,240 nests (Fig. 5.7, right). Yet the swift tower ended up being a sparrow tower. The problem is that the hole size of the nest box is

Fig. 5.6 Holes for the nesting of swifts and bats have been incorporated into the facade of a building in Germany during renovation

Fig. 5.7 Examples of urban sculptures and structures using the preference of birds to nest in holes (Left: A sculpture in Jing'an Sculpture Park, Shanghai attracts a large number of sparrows to build their nests; Right: The swift tower in Beijing Olympic Park)

too small and the density is too high, greatly increasing the inter-species competition among swifts in the case of limited food and making swifts unable to adapt to such artificial nests. Therefore, in the design of hole space, such factors as the nesting space, the feeding space, and the population density of target species need to be carefully considered.

5.5 Key Planning and Design Indicators for Improving Biodiversity Performance of Urban Built Environment

According to the affecting factor variables summarized in Sect. 5.1, along with the three spatial levels of urban master plan, regulatory plan, and urban design, key planning and design indicators based on the biodiversity performance of the urban built environment are extracted as a supplement to the design strategies and further requirements for regulation. The indicators should be extracted in accordance with the following principles:

(1) Take the operability and manageability of planning and design as the primary principle, convert the affecting factor variables into control and guidance indicators for planning and design, some into indicators with more understandable names, and some into regulable indicators with core components from the factors;
(2) Combine the appropriate amount of relevant affecting factors with similar effects into one indicator;
(3) Divide the indicators into upgrading indicators and new indicators according to the existing planning and design index system.

Accordingly, six optimization indicators are extracted in the study: one for the master plan level, two for the regulatory level, and three for the urban design level (Table 5.4).

M1-Area of major wildlife habitats (hm^2)

Definition: Sect. 3.4.1 expounds on the importance of designating major wildlife habitats. According to the theory of conservation ecology, all species need habitats with a certain scale to survive. Take birds as an example, only large-scale green spaces with good planting can accommodate dense-forest, forest, and forest-margin birds at the same time, fulfilling the purpose of biodiversity. Therefore, the area of a habitat must reach a certain level to meet the most fundamental ecological requirement for wildlife.

Setting method: The indicator should be set according to the habitat requirements of a certain target species population. For example, the indicator of a woodland habitat can be set according to the habitat requirements of bird population.

R1-Ratio of green infrastructure area (%)

Definition: Green infrastructure is a concept inspired by gray infrastructure (such as roads, sewers, and utility lines) or social infrastructure (such as schools, hospitals, and prisons). It is essentially a natural system on which urban sustainable development depends, and the foundation for cities and their residents to obtain sustainable natural services (An and Shen 2013). It is generally considered to play an important supporting role in biodiversity protection and improvement. Green infrastructure is composed of multifunctional open spaces. At the regulatory level, artificial facilities

Table 5.4 Planning and design optimization indicators for improving the biodiversity performance of the urban built environment

Level	No.	Name	Unit	Type	Planning and design principle	Affecting factor	Effect
Master Plan	M1	Area of major wildlife habitats ↑	hm^2	Upgrading	Quantity assurance	Area of major wildlife habitats	Area effect
Regulatory Plan	R1	Ratio of green infrastructure area ↑	%	Upgrading	Size increase	Green space rate, water surface rate, building density	Ratio effect
	R2	Area of the largest concentrated green space ↑	m^2	New	Greening centralization	Area of the largest patch of green space	Area effect
Urban Design	D1	Coverage area ratio of trees in green space ↑	%	New	Benefit improvement	Coverage area of trees	Area effect
	D2	Average height of trees ↑	m	New	Planting suitability	Tree height	Height effect
	D3	Setback rate of woods' edge in green space↓	%	New	Disturbance reduction	Setback rate of woods' edge	Edge effect

Note "↑" means lower limit indicator under regulation; "↓" means upper limit indicator under regulation

can be regarded as green infrastructure if they can provide some natural life support, including parks, community green spaces, roadside green spaces, water bodies and waterfront green spaces, green corridors, ponds, rain gardens, footpaths, and some natural reserve areas such as wetland and woodland. Ratio of green infrastructure area (%) = the sum of all green infrastructure areas/land area.

In conventional planning, the green space ratio, green coverage ratio, and the per capita area of public green space are generally used as the regulatory indicators of green space, focusing on providing people with green services to cultivate their mind and body; while green infrastructure lays more emphasis on the nature protection and wildlife habitat supporting service of green space. When more attention is paid to the artificial landscape and the natural ecological process is ignored, the ratio of green infrastructure area will be lower than the rate of green space; when all the elements in green space have the capability of providing natural ecological support, the ratio of green infrastructure area will surely be higher than the rate of green space. By upgrading the "rate of green space" to the "ratio of green infrastructure area," we emphasize the natural ecological function of green space, namely to support the inhabitation of organisms.

Setting method: Ratio of green infrastructure area > Green space ratio.

R2-Area of the largest concentrated green space (m^2)

Definition: The larger the area of the largest concentrated green space is, the more favorable it is for the stability of biological populations. It is in essence the rationalization of green space structure. According to the practical value of planning and management, the area of the largest concentrated green space is mainly for the regulation of residential green space.

Setting method: The indicator should be set according to the biodiversity target of the project. According to the survey data in Chap. 4, except for such urban green spaces as the small wetland beside Shanghai Science and Technology Museum, Lujiazui Central Greenland, Mingzhu Park, Douxiang Park, and the woods near the Shanghai Science and Technology Museum Station on Metro Line 2, the average species richness of the blocks is 8, and the average Shannon-Wiener diversity index is 1.3. Those can be used as the basic diversity targets for birds on the street scale, i.e., the species number of birds that can be seen is more than twice the number of the four most common birds in cities—Eurasian Tree Sparrow, Light-vented Bulbul, Chinese Blackbird, and Spotted Dove. As ten dominant species and common species are found in the survey, those with a species richness of 10 and a Shannon-Wiener diversity index of 1.5 will be considered to have high biodiversity.

As shown in Sect. 4.5.2, the area of the largest patch of green space is correlated with the species richness and the Shannon-Wiener diversity index (p-value = 0.01), with a correlation coefficient of 0.690 (strongly correlated) and 0.515 (moderately correlated), respectively. Polynomial regression analysis has been done on the area of the largest patch of green space, the species richness, and the Shannon-Wiener diversity index (with R^2 changed to greater than 0.5, and the goodness-of-fit basically meeting the requirements). According to the fitted curve, if the species richness is set to be 8 and the diversity index 1.3, then the area of the largest green space (the area of the maximum green space patch) should be no less than 3,000 m^2; if the species richness is set to be 10 and the diversity index 1.5 for a higher standard, then the area of the largest concentrated green space should be no less than 4,000 m^2 (Figs. 5.8 and 5.9).

According to the *Regulation of Shanghai Municipality on Greening* (2015), the proportion of the green space area in a newly built residential community to its total area shall not be less than 35%, and the area of land used for development of concentrated green space shall not be less than 10% of its total area. If the total area of a residential community is 5 ~ 10 hm^2, then the total area of the concentrated green space should be no less than 5,000 ~ 10,000 m^2. Therefore, the area of the largest concentrated green space should be no less than 3,000 ~ 4,000 m^2, which is in line with the current regulations.

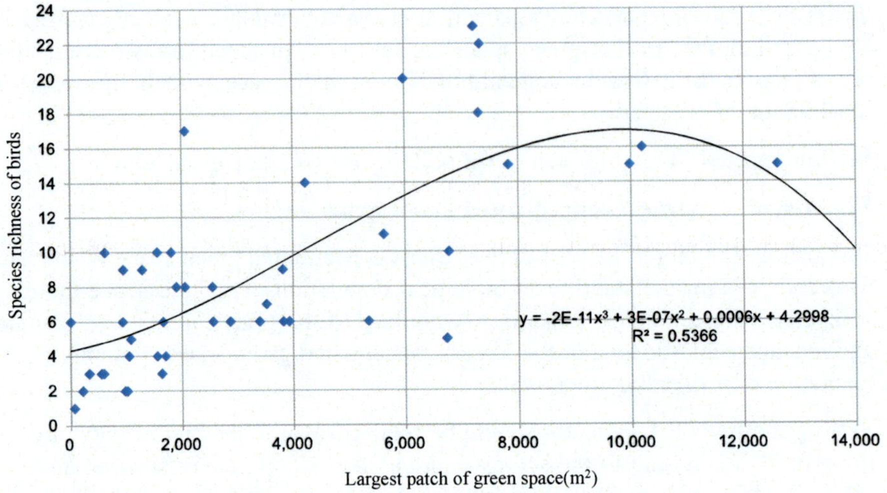

Fig. 5.8 Fitted curve of the area of the largest patch of green space in blocks along the Century Avenue and the species richness of birds

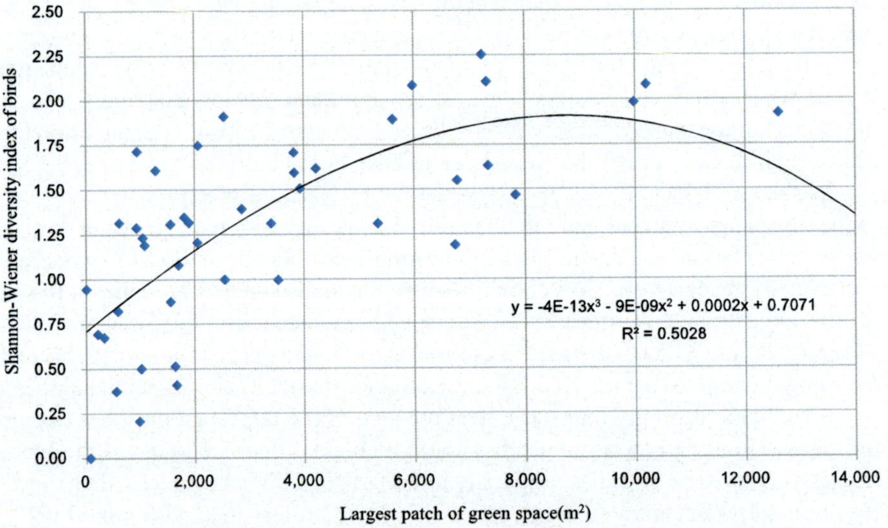

Fig. 5.9 Fitted curve of the area of the largest patch of green space in blocks along the Century Avenue and the Shannon-Wiener diversity index of birds

D1-Coverage area ratio of trees in green space (%)

Definition: Coverage area ratio of trees refers to the proportion of the trees-planted area into the total green area. It reflects the proportion of trees that can provide more spatial niches for organisms to feed, nest, and rest, demonstrating the soundness of vegetation structure.

Setting method: As shown in Sect. 4.5.3, the percentage of land covered by trees and the species richness are significantly correlated with and the Shannon-Wiener diversity index (p-value-0.01), with a correlation coefficient of 0.656 (strongly correlated) and 0.516 (moderately correlated), respectively. Based on the field survey data in Chapter 5, the author prepares a scatter diagram of the percentage of land covered by trees in the sample block, the species richness, and the Shannon-Wiener diversity index. As the polynomial regression goodness-of-fit is less than 0.3, a fitted curve cannot be used to define the afforestation rate for different biodiversity targets. However, by studying the scatter diagram, it can be found that when the species richness exceeds 8 and the diversity index exceeds 1.3, the afforestation rate must be above 20%; when the species richness exceeds 10 and the diversity index exceeds 1.50, the afforestation rate must be above 30% (Figs. 5.10 and 5.11).

D2-Average height of trees (m)

Definition: Tree height refers to the vertical distance between the rhizomes of a tree above the ground and the top of the canopy. The average height of trees reflects the height of spatial niches provided by vegetation overlay for birds and other organisms. The higher the average height of trees, the more vegetated vertical niches for birds. This indicator can provide basis for the selection of tree species.

Setting method: The indicator is set according to the height required for the tree layer of the target species. As shown in Sect. 4.6.3, the average height of trees is

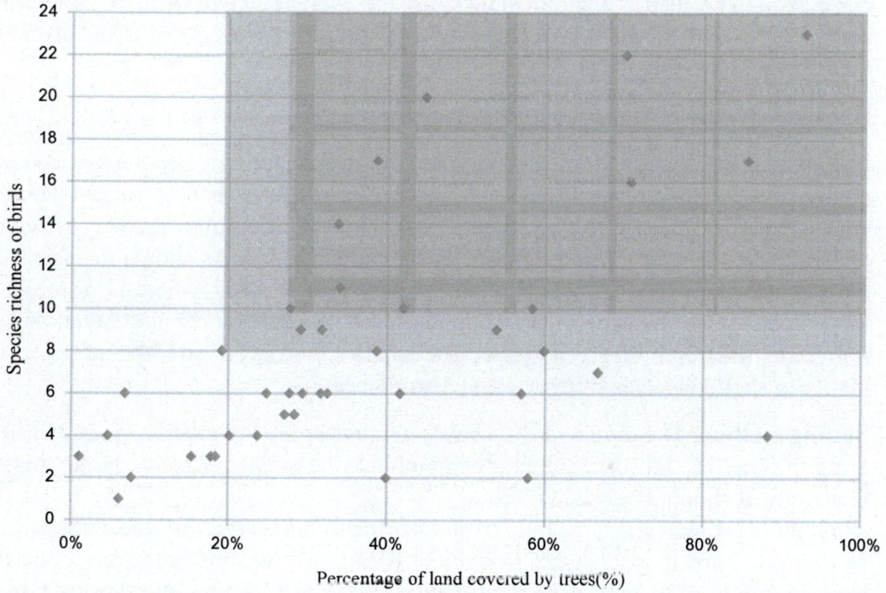

Fig. 5.10 Scatter diagram of the percentage of land covered by trees in blocks along the Century Avenue and the species richness of birds

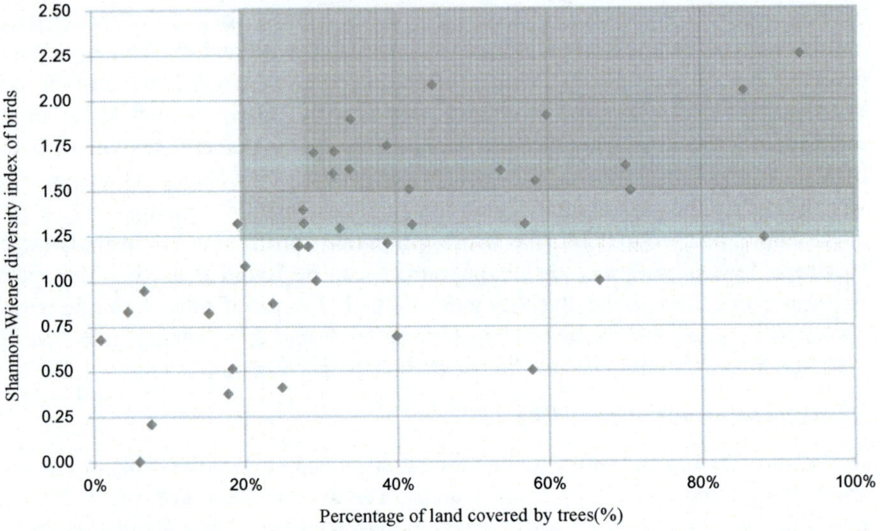

Fig. 5.11 Scatter diagram of the percentage of land covered by trees in blocks along the Century Avenue and the Shannon-Wiener diversity index of birds

significantly correlated with species richness and diversity index (p-value = 0.01), with a correlation coefficient of 0.480 and 0.520, respectively, indicating a moderate correlation. The scatter diagram shows that the average height of trees falls in the sections of 4–6, 6–8, and 8–10 m, the average values of species richness and diversity index keep increasing (Figs. 5.12 and 5.13).

D3-Setback rate of woods' edge in green space (%)

Definition: Specifically, it refers to the ratio of the distance between the first layer of continuous tree surface in a roadside green space close to the pedestrian path and the edge of the pedestrian path to the depth of the green space, reflecting the resistance of the green space to the interference of human activities. As shown in Fig. 5.14, the setback rate of woods' edge in green space (%) = *Dt/D*, where *Dt* represents the border retreat distance of the continuous tree surface on the first layer, and *D* represents the depth of green space. The smaller the retreat rate, the stronger the ability to shelter species from external disturbance.

Setting method: The retreat rate of forests in green space is a relative index. Considering the line of sight and the landscape effect, the smaller the rate is, the better. Generally, it should not exceed 0.5.

Of these six indicators, the ratio of green infrastructure area, the area of the largest concentrated green space, the Coverage area ratio of trees in green space, and the average height of trees can be incorporated into the planning and design system as general ecological control and guidance indicators because they can, apart from improving the biodiversity performance of the built environment, retain and purify rain water, purify the air, and regulate the micro climate.

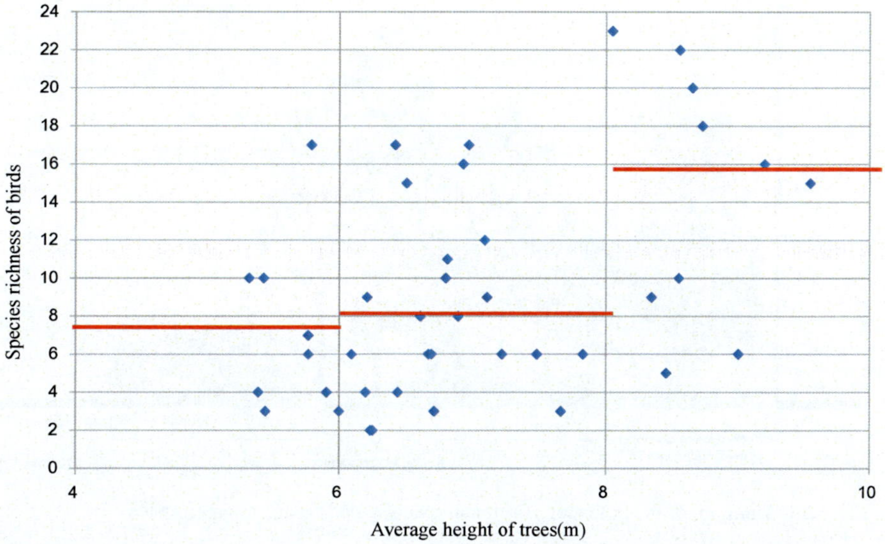

Fig. 5.12 Scatter diagram of the average height of trees in blocks along the Century Avenue and the species richness of birds (X-axis: average height of trees; Y-axis: species richness of birds; and the red line is the average value of each section)

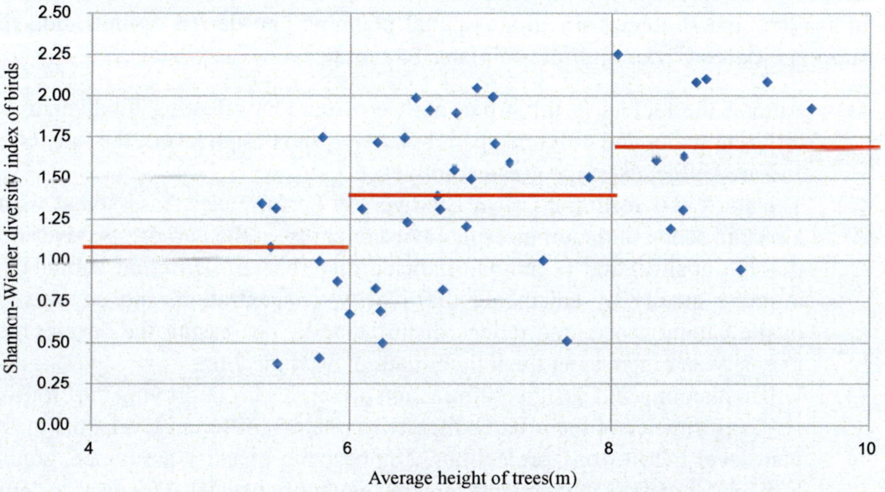

Fig. 5.13 Scatter diagram of the average height of trees in blocks along the Century Avenue and the Shannon-Wiener diversity index of birds (The red line is the average value of each section)

Fig. 5.14 Diagram of the indicator of setback rate of woods' edge in green space

5.6 Summary

Based on theoretical and empirical studies, this chapter discusses the principles, indicators, and strategies for urban spatial planning and design optimization that support biodiversity of multiple habitats. To sum up:

(1) Among the factors in the urban built environment affecting biodiversity on different scales and different spatial planning and design levels, the ratio effect and area effect exert the greatest influence.

(2) The study sets forth the Urban Biodiversity Performance Assessment within a certain scope of the urban built environment from the two perspectives of Q (habitat quality) and D (human-induced disturbance). Different habitats can improve quality by enhancing the density, concentration, and connectivity of the habitat space, and reduce disturbance by decreasing the development intensity and increasing the human-nature disconnection.

(3) Urban planning and design optimization principles for improving the biodiversity performance of the urban built environment at different levels: the master plan level focuses on "protection," emphasizing quantity assurance, zoning, level classification, and networking; the regulatory plan level focuses on "guidance and control," emphasizing size increase, greening centralization, distance control, and corridor access; and the urban design level focuses on "promotion," emphasizing benefit improvement, planting suitability, disturbance reduction, and symbiosis.

(4) Urban planning and design optimization strategies for improving the biodiversity performance of the urban built environment at different levels:

at the master plan level: Ascertain a minimum threshold space of ecological land and define the detailed proportion of each ecological land category; establish a multidimensional network for the protection of multiple habitats: "nature reserves-major habitats-supplementary habitats-symbiotic habitats,"; moderately control the intensity of development in suburban areas to avoid encroachment of development on key natural habitats, and improve the connectivity of urban ecological river network;

at the regulatory plan level: use the "sink-corridor-stepping stone" habitat chain system to create the green space pattern in semi-artificial habitats; encourage ecological development of artificial water surfaces; conduct afforestation within a certain range of the riverside area; and highlight the value of natural assets in urban renewal on abandoned land after natural succession;

at the urban design level: optimize the shape and spacing of centralized green spaces; promote the multi-layered tree-shrub-herb planting structure and the foodborne native vegetation; use layered and three-dimensional greening as off-ground habitat patches and corridors; and consider the preference of birds to nest in holes in the design of urban buildings, structures, and street furniture.

(5) Key planning and design indicators for improving the biodiversity performance of the urban built environment at different levels. The indicators at the master plan level include: M1-area of major wildlife habitats (hm^2); the indicators at the regulatory plan level include: R1-area ratio of green infrastructure (%), R2-area of the largest concentrated green space (m^2); indicators at the urban design level include: D1-coverage area ratio of trees in green space (%); D2-average height of trees (m), D3- setback rate of woods' edge in green space (%).

References

An C, Shen QJ (2013) Method of Constructing Network of Green Infrastructure Based on Ecological Performance of Space Utilization. Landsc Archite 2:22–31 (in Chinese)

Brenneisen S (2006) Space for urban wildlife: designing green roofs as habitats in Switzerland. Urban Habit 4(1):27–36

Fernández-Juricic E, Jokimaki J (2001) A habitat island approach to conserving birds in urban landscapes: case studies from southern and northern Europe. Biodivers Conserv 10:2023–2043 (in Chinese)

Francis RA (2011) Wall ecology a frontier for urban biodiversity and ecological engineering. Progress Phys Geogr 35(1):43–63

Hao RM, Zhang MJ (2015) Noteworthy points in the compilation of conservation planning for urban biodiversity. Chinese Landscape Archit 31(8):5–9 (in Chinese)

Li J, Tu QY (2014) Improvement of urban space in super cities from the perspective of ecological civilization—take shanghai as an example. Shanghai Urban Manag 6:9–14 (in Chinese)

Lin XD (2001) Urban and Rural Ecology. Chan's Arch-Publishing Co, Ltd, Taipei, p 14 (in Chinese)

Lin XD (2007) Green building-eco, energy saving, waste reduction and healthy. China Architecture Publishing & Media Co, Ltd, Beijing, pp 84–94 (in Chinese)

Shanghai Environmental Protection Bureau (2013) Shanghai greening and city appearance Administration, Shanghai Municipal Agriculture Commission, Shanghai Water Affairs Bureau (2013) Shanghai biodiversity conservation strategy and action plan (2012–2030)

Tang CD, Guan LQ, Xie YM (2003) Current situation and thinking of wildlife conservation and management in Shanghai. Wildlife 6:51–53 (in Chinese)

Tong XP, Zhou L, Yang PP (2009) Foodborne native plants for birds and their application. In: Proceedings of the annual academic conference of China Botanical Garden. Nanning, China (in Chinese)

Werner P, Groklos M, Eppler G et al (2016) Schutz gebäudebewohnender tierarten vor dem hintergrund energetischer gebäudesanierung. Hintergründe, Argumente, Positionen. in Städten und Gemeinden. Technical Report for BFN. https://www.bfn.de/fileadmin/BfN/planung/siedlung/Dokumente/Gebaeudebruetende_Tierarten_2016_-_Positionspapier.pdf. Accessed 1 Sept 2016

Yan WT, Xiao JH, Hu H et al (2012) Urban spatial structure and environmental performance: review and thought. Urban Plann Forum 2:50–59 (in Chinese)

Chapter 6
Conclusions and Prospects

6.1 Research Conclusion

In response to the encroachment of high-density urbanization on natural biotic habitat space and the lack of a biodiversity concern in the ecological urban planning, this book, by focusing on the goal of harmonious coexistence of organisms and people in cities, explores the demand of urban organisms for habitat space, examines, from the perspective of multidimensional multiple habitats and through empirical analysis, the effects and influence mechanism of the urban built environment on biodiversity on different scales, and proposes the regulatory elements in urban space planning and design to effectively protect biodiversity and improve service functions in the ecosystem, thereby laying the groundwork for addressing the contradiction between the development of "high-density and compact" urban space and the need to preserve "resilient" living space on biotic habitats as China is becoming increasingly urbanized. Below are the main conclusions.

6.1.1 Theoretical Development: Relationship Between Urban Biodiversity and Built Environment

(1) Urban multiple overlapping habitats

This book sets forth the theory of urban multiple overlapping habitats from the perspective of multidimensional overlapping between man and nature, proposes five types of urban multiple habitats and describes their functions and supply potential. According to the nature of urban land and the degree of human-nature overlapping, the five types of habitats are defined as follows: near-natural habitats on the non-development land, semi-natural habitats in the urban parks and public green spaces, semi-artificial habitats in the attached green spaces, artificial habitats on the sealed surfaces, and habitats with natural-succession on the abandoned land.

J. Gan, *Urban Biodiversity and Built Environment*,
https://doi.org/10.1007/978-981-16-0949-7_6

(2) **The relationship between urban biodiversity and built environment**

This book puts forward the theory of correlation between urban biodiversity and built environment. The affecting factors of the "dual perspective" include the habitat quality factors and the human-induced disturbance factors, and the six effects are: the ratio effect, the area effect, the edge effect, the distance effect, the network effect, and the height effect. An interconnected action mechanism is created through various affecting factors on multiple habitats of different scales.

6.1.2 *Influence Mechanism: The Influence Mechanism of Built Environment on Biodiversity on Two Scales*

Based on the empirical research in Shanghai, this book discusses the influence mechanism of three built environment variables (development intensity, ecological land, and vegetation) on biodiversity on the macro scale and on the meso-micro scale.

(1) **The influence mechanism of macro-scale urban built environment on biodiversity**

Influence of development intensity: As different forms of human-induced disturbance, the economic development intensity (GDP per km^2), the population concentration (population density), and the intensity of urban development (the building density, the road network density, the high-rise building area and its ratio) have significantly negative impact on the overall biodiversity index and the species richness, such as habitat loss of primary organisms, and have a greater effect on organisms sensitive to environmental changes such as mammals and fish than on other species.

Influence of ecological land: The size of various types of ecological land directly affects the biodiversity habitat quality. For Shanghai, the size of water area and wetland has the greatest influence on biodiversity, followed by that of cultivated land. Therefore, they should be given priority in the regulation of ecological space. To ensure the stability of populations, the development of both green spaces in parks as semi-natural habitats and major wildlife habitats as near-natural habitats should guarantee the area effect of habitats.

In addition to the influence mechanism of spatial morphological variables (such as the density of ecological land patches, and the landscape combination and aggregation indexes) on biodiversity that has been verified in other relevant studies, this study finds that the improvement of the closure, line-node rate, and network connectivity of the urban water network can promote the conduit effect of the ecological network, and increase the inhabitation, migration, and gene exchange of birds, fish, amphibians, and reptiles on the macro scale, but blocks those of mammals and invasive species.

Combined influence of development intensity and ecological land use on major wildlife habitats: The study finds that the incorporation of different levels of wildlife habitats into urban statutory planning is an effective means to effectively safeguard multiple habitat spaces for wildlife on the macro scale. The habitat quality of exurbs is better than that of suburbs and city boundaries. The size of woodland and lake reservoir habitats, the size of habitats in suburbs and exurbs, and the urban development in the surrounding area have some effects on the habitat quality. Therefore, when demarcating animal wildlife habitats, we should pay special attention to the residual semi-natural habitats in the suburban areas, ensure the area of habitats and avoid the interference of the surrounding urban development.

(2) **The influence mechanism of urban built environment on diversity performance of birds on the meso-micro scale**

The bird community structure in high-density urban centers is singular, with resident birds in the majority. The wetland habitats with natural-succession on the abandoned land can serve as important "sink" habitats for urban wild birds as they boast greater species richness of birds, higher diversity, and more complex bird community structure than the semi-natural habitats in the urban parks and public green spaces and the semi-artificial habitats in the attached green spaces. The species richness of birds in the semi-artificial habitats in the attached green spaces is lower than that of the semi-natural habitats in the urban parks and public green spaces, while the species richness of residential land is higher than that of public facilities. The bird diversity of high-rise residential areas with centralized green spaces is better than that of multi-storey residential communities.

Influence of development intensity: The FAR, the building density, and the ratio of green space and water surface, which are development intensity variables deciding the overlapping relationship between human settlements and biotic habitats, have a significant effect on avian diversity on the meso-micro scale. The building density has the largest effect on bird diversity. Specifically, in the development of high-density urban areas, high-rise and low-density buildings are more favorable for bird diversity than low-rise and high-density buildings. Therefore, reducing the area of buildings and providing more substrate space for organisms to inhabit is an effective way to improve the biodiversity of urban central areas.

Influence of ecological land: Both the size of green space patches (the total area of green space patches, the largest patch area of green space, and the patch density) and their spatial form (the perimeter area ratio of green space (m/m^2), the perimeter area ratio of the largest patch of green space, the average proximity index, the landscape aggregation index, the landscape combination index) have a great effect on the diversity of birds: the former can offer carrying substrates for bird communities to inhabit, and the latter can improve the distance effect and the networking effect of habitats while reducing the edge interference, thus improving the overall quality of habitats.

The spatial morphological effect of waters is more obvious in residential communities and public service facilities.

Influence of vegetation: Mainly including vegetation size (coverage area of trees/ground cover), vegetation structure (the species number of trees/ground cover plants), vegetation form (the average height of trees, the ratio of disturbance boundary in roadside green spaces, and the border retreat rate of first-floor trees). The tree layer and the ground layer are the main feeding and nesting spatial niches for bird communities, which has also been verified in the actual analysis of microhabitats. The vegetation size reflects the area effect of bird carrying substrates, and the vegetation species and the average height of trees reflect the attraction of multi-layered planting structures to birds and the potential of birds to share overlapping spaces with humans in high-density urban areas. The interfered/proximate border ratio and the setback rate of woods' edge reflect the level of roadside green spaces as "stepping stone" habitats to resist external disturbance pressure.

In different function areas, the effect of variables in the shrub layer of park green space blocks has improved.

Overall influence: Results show that the building density and the average height of trees are variables which have the greatest effect on the diversity of birds. When the development intensity is identified, the patch cohesion index of green space, the average height of trees, the species richness of groundcover plants, and the percentage of land covered by groundcover plants are the four major variables affecting the diversity of birds. The patch cohesion index of green space, exerts the greatest influence, followed by the average height of trees. This can serve as a reference for the arrangement of green space and the spatial distribution of vegetation.

Other influence: The actual analysis of microhabitats also finds that a large number of dominant species and common species of resident birds stay on buildings/structures for resting, even foraging and nesting, which is a unique phenomenon of the urban built environment, and preliminarily validates the possibility of urban wildlife using vertical off-ground habitats.

6.1.3 Planning Response: Planning and Design Optimization to Improve Biodiversity Performance of Urban Built Environment

Based on the key affecting factors extracted from empirical study, this book focuses on the biodiversity performance of the built environment in multiple habitats, **proposes the Q/D (Quality/Disturbance) Urban Biodiversity** Performance **Assessment of the built environment, and identifies the planning and design regulation objectives for multiple habitats**.

Based on the existing system of urban planning and design, this book **proposes the following principles for optimization of spatial planning and design: "quantity assurance, zoning, level classification, and networking" on the master plan level,**

"size increase, greening centralization, distance control, corridor access" on the regulatory plan level, and "benefit improvement, planting suitability, disturbance reduction, symbiosis" on the urban design level to improve the biodiversity performance of the urban built environment, and develops an optimization strategy package consisting of 12 strategies and six key indicators.

6.2 Prospects

Like humans, organisms are also residents of cities. They are of inestimable value to the stability of urban ecosystems and are entitled to urban space just like humans do. The reasons underlying the selection preferences of biotic habitat space, however, are quite complex. This book, from the perspective of urban space planning and design, tries to explore in a thorough manner the factors in the built environment affecting the biodiversity of multiple habitats in cities, and then proposes the principles, indicators, and strategies for optimization by taking into consideration the current planning and design system, thus providing references for the all-round construction ecological cities. Of course, this topic entails more in-depth research in the future.

(1) **From birds to other species—on the effect and mechanism of urban built environment on biodiversity of different populations**

Due to limited research time, this book only chooses birds on the meso-micro scale as the indicator species to investigate and analyze the influence of the built environment on biodiversity. Urban biological communities on the trophic level are abundant, so by studying the effects of the urban built environment on the biodiversity of different populations, we can gain a deeper understanding of the relationship and difference between species with regard to the environmental effects on biodiversity, and further improve the theoretical framework of effects. For example, as the most diverse and abundant biological community in cities, insects play an important role. Like birds, they are often used as the key species to indicate the changes of urban habitats. Therefore, research on the diversity of insects is also of great significance to the monitoring and protection of urban biodiversity.

(2) **From "correlation" to "symbiosis"—on the harmonious relationship between urban biodiversity and built environment**

According to the laws of nature, the exchange of demands among various organisms maintains the equilibrium of the environment where they keep each other in check while promoting the ecological balance. The total amount of living space and resources on earth determines the size and range of activities of organisms. In nature, all kinds of creatures observe certain rules, which results in proper and orderly allocations of living space and resources. In cities, human activities are bound to encroach on the habitats of primary organisms, hence a new symbiotic relationship of dynamic balance with urban organisms. Such relationship is neither antagonistic nor monopolistic. Urban biodiversity, of course, has its downsides. Therefore, in urban development, we should reserve an appropriate amount of living space for wildlife, but should never encourage excessive reproduction of creatures to cause another type of ecological imbalance. This book mainly discusses the positive and negative effects of the relationship between urban biodiversity and the built environment, follow-up study can further explore the harmonious relationship between people and different kinds of organisms in the urban built environment, i.e., how the harmonious and symbiotic relationship between the diversity of different species and the built environment varies in different stages of urban development and under different development goals.

(3) **From "biodiverse city" to "biophilic city"—on the planning and design approaches of human-nature interactive space**

Biodiversity plays a key role in maintaining urban ecological security and ecological balance and in improving human settlements. This book reexamines the lack of spatial planning and design in ecological cities from the perspective of biodiversity. "People" mainly influence biodiversity through the intervention of development activities on the built environment. In fact, humans have an instinct to "the urge to affiliate with other forms of life," an innate interest and curiosity that psychologists and biologists call "biophilia" (Wilson 1984). The positive contact and interaction between humans and organisms are not only beneficial to the physical and mental health of humans, but also prompts people to take initiative in protecting and promoting biodiversity. Therefore, in the subsequent studies, the perspective of biodiversity can be elevated to that of biophilia, which features the interaction between humans and organisms, to further explore how to encourage citizens in space planning and design to use vision, hearing, touch and other senses to experience the charm of nature and obtain a sense of happiness and intimacy, such as joy, relief, comfort, and inspiration, thereby promoting the development of urban environment where people interact with organisms.

(4) **From the ground to off-ground—the biodiversity of off-ground habitats and the planning and design approaches**

This book preliminarily demonstrates the potential of three-dimensional surface substrates to provide habitats for urban wild birds, but due to the limited energy and time and the current conditions of substrates, there is no further observation and analysis of three-dimensional off-ground habitats. For high-density Chinese cities, if biodiversity is incorporated into the three-dimensional space greening development, more space will be available for urban organisms, which will also be a supplement to the urban multidimensional multiple habitat system.

(5) From theory to practice—guidelines on and demonstration of the planning and design of high-density urban space supporting biodiversity in multiple habitats

Based on theoretical and empirical findings, this book identifies the factors in the built environment affecting biodiversity on different dimensions, and proposes some principles, indicators, and strategies for optimizing spatial planning and design. Those results will be applied systematically in different types of cases in the future, where guidelines on ecological planning and design at different levels will be developed to further verify the applicability of the research results. The design effectiveness can be examined through demonstration projects, which can promote the implementation and application of biodiversity-oriented urban planning.

Reference

Wilson EO (1984) Biophilia. Harvard University Press, Cambridge, UK

Appendix
Survey Results of Birds in the Study Area of Century Avenue in Shanghai

J. Gan, *Urban Biodiversity and Built Environment*,
https://doi.org/10.1007/978-981-16-0949-7

No.	Order	Family	Bird Species	Resident and migrant types	Protection level	Encounter rate(P_i)				
						Spring	Summer	Autumn	Winter	Whole year
1	I.PODICIPEDIFORMES	1. Podicipedidae (Grebes)	Little Grebe (*Tachybaptus ruficollis*)	R	#	+	+	-	-	+
2	II.PELECANIFORMES	2. Ardeidae (Herons, Egrets, Bitterns)	Little Egret (*Egretta garzetta*)	S	#	+	+	++	+	+
3	II.PELECANIFORMES	2. Ardeidae (Herons, Egrets, Bitterns)	Chinese Pond Heron (*Ardeola bacchus*)	S	#	–	–	–	+	+
4	II.PELECANIFORMES	2. Ardeidae (Herons, Egrets, Bitterns)	Black-crowned Night Heron (*Nycticorax nycticorax*)	R	#	++	+++	++	++	++
5	III.GRUIFORMES	3. Rallidae (Rails, Crakes, Coots)	Common Moorhen (*Gallinula chloropus*)	R	#	+	+	+	+	+
6	IV.CHARADRIIFORMES	4. Laridae (Gulls, Terns, Skimmers)	Mew Gull (*Larus canus*)	W	#	++	–	–	++	++
7	V.COLUMBIFORMES	5. Columbidae (Doves, Pigeons)	Spotted Dove (*Streptopelia chinensis*)	R	#	++	+	+++	++	++
8	VI.CORACIIFORMES	6. Alcedinidae (Kingfishers)	Common Kingfisher(*Alcedo atthis*)	R	#	+	+	+	+	+
9	VII.BUCEROTIFORMES	7. Upupidae(Hoopoes)	Common Hoopoe(*Upupa epops*)	P/S	#	+	–	–	+	+
10	VIII.PASSERIFORMES	8. Hirundinidae (Martins and Swallows)	*Barn Swallow (Hirundo rustica)*	S	#	+	+	–	–	+
11	VIII.PASSERIFORMES	9. Motacillidae (Wagtails, Pipits)	White Wagtail (*Motacilla alba*)	P/W	#	+	+	++	++	++
12	VIII.PASSERIFORMES	9. Motacillidae (Wagtails, Pipits)	Olive-backed Pipit (*Anthus hodgsoni*)	W	#	+	+	–	+	+
13	VIII.PASSERIFORMES	10. Pycnonotidae (Bulbuls)	Light-vented Bulbul (*Pycnonotus sinensis*)	R	◇#	+++	+++	+++	+++	+++
14	VIII.PASSERIFORMES	10. Pycnonotidae (Bulbuls)	Brown-eared Bulbul (*Hypsipetes amaurotis*)	P		+	–	–	+	+
15	VIII.PASSERIFORMES	11. Laniidae (Shrikes)	Brown Shrike(*Lanius cristatus*)	S	#	–	–	+	–	+
16	VIII.PASSERIFORMES	11. Laniidae (Shrikes)	Long-tailed Shrike (*Lanius schach*)	R	◇/#	++	+	++	+	++

(continued)

(continued)

No.	Order	Family	Bird Species	Resident and migrant types	Protection level	Encounter rate(P_i)				
						Spring	Summer	Autumn	Winter	Whole year
17	VIII.PASSERIFORMES	12. Sturnidae (Starlings)	Silky Starling (*Spodiopsar sericeus*)	R	#	+	–	+	+	+
18	VIII.PASSERIFORMES	12. Sturnidae (Starlings)	Crested Myna (*Acridotheres cristatellus*)	R	◇#	+	+	+	+	+
19	VIII.PASSERIFORMES	13. Corvidae (Crows, Jays)	Azure-winged Magpie (*Cyanopica cyanus*)	R	#	++	+	++	++	++
20	VIII.PASSERIFORMES	13.Corvidae (Crows, Jays)	Common Magpie (*Pica pica*)	R	◇/#	+	+	–	+	+
21	VIII. PASSERIFORMES	14. Turdidae (Thrushes)	*White's Thrush (Zoothera aurea)*	P	#	+	+	+	+	+
22	VIII. PASSERIFORMES	14. Turdidae (Thrushes)	Grey-backed Thrush (*Turdus hortulorum*)	P	#	+	–	–	+	+
23	VIII. PASSERIFORMES	14. Turdidae (Thrushes)	Chinese Blackbird (*Turdus mandarinus*)	R	◇	+++	+++	+++	+++	+++
24	VIII.PASSERIFORMES	14. Turdidae (Thrushes)	*Eyebrowed Thrush* (*Turdus obscurus*)	P		-	-	+	+	+
25	VIII.PASSERIFORMES	14. Turdidae (Thrushes)	Pale Thrush (*Turdus pallidus*)	W/P	#	+	–	+	++	+
26	VIII.PASSERIFORMES	14. Turdidae (Thrushes)	Dusky Thrush (*Turdus eunomus*)	W	#	++	–	–	++	+
27	VIII.PASSERIFORMES	15. Muscicapidae (Old World Flycatchers, Robins, Redstarts)	Orange-flanked Bluetail (*Tarsiger cyanurus*)	W	#	–	–	+	+	+
28	VIII. PASSERIFORMES	15. Muscicapidae (Old World Flycatchers, Robins, Redstarts)	Oriental Magpie Robin (*Copsychus saularis*)	R	#	+	-	+	+	+
29	VIII.PASSERIFORMES	15.Muscicapidae(Old World Flycatchers, Robins, Redstarts)	Daurian Redstart (*Phoenicurus auroreus*)	W	#	+	+	+	+	+
30	VIII.PASSERIFORMES	15. Muscicapidae (Old World Flycatchers, Robins, Redstarts)	Dark-sided Flycatcher (*Muscicapa sibirica*)	P	#	+	-	+	-	+

(continued)

(continued)

No.	Order	Family	Bird Species	Resident and migrant types	Protection level	Encounter rate(P_i)				
						Spring	Summer	Autumn	Winter	Whole year
31	VIII.PASSERIFORMES	15. Muscicapidae(Old World Flycatchers, Robins, Redstarts)	Asian Brown Flycatcher (*Muscicapa dauurica*)	P	#	+	-	+	+	+
32	VIII. PASSERIFORMES	15. Muscicapidae (Old World Flycatchers, Robins, Redstarts)	Narcissus Flycatcher (*Ficedula narcissina*)	P	#	+	-	-	-	+
33	VIII. PASSERIFORMES	15. Muscicapidae (Old World Flycatchers, Robins, Redstarts)	Mugimaki Flycatcher (*Ficedula mugimaki*)	P	#	-	-	+	+	+
34	VIII. PASSERIFORMES	16. Sylviidae (Old World Warblers and Parrotbills)	Vinous-throated Parrotbill (*Sinosuthora webbiana*)	R		++	+	++	++	++
35	VIII. PASSERIFORMES	17. Cisticolidae (Cisticolas)	Plain Prinia (*Prinia inornata*)	R		–	+	–	–	+
36	VIII. PASSERIFORMES	18. Phylloscopidae (Leaf-warblers)	Pallas's Leaf Warbler (*Phylloscopus proregulus*)	P	#	+	+	+	+	+
37	VIII. PASSERIFORMES	18. Phylloscopidae (Leaf-warblers)	Yellow-browed Warbler (*Phylloscopus inornatus*)	P	#	+	+	+	+	+
38	VIII. PASSERIFORMES	18. Phylloscopidae (Leaf-warblers)	Eastern Crowned Warbler (*Phylloscopus coronatus*)	P	#	+	–	–	–	+
39	VIII. PASSERIFORMES	19. Remizidae(Penduline Tits)	Chinese Penduline Tit (*Remiz consobrinus*)	W	#	+	–	–	–	+
40	VIII. PASSERIFORMES	20.Aegithalidae (Long-tailed Tits)	Black-throated Bushtit (*Aegithalos concinnus*)	R	#	–	–	+	–	+
41	VIII. PASSERIFORMES	21. Paridae(Tits)	Cinereous Tit (*Parus cinereus*)	R	◇#	+	+	+	+	+
42	VIII. PASSERIFORMES	22. Passeridae (Old World Sparrows)	Eurasian Tree Sparrow (*Passer montanus*)	R	#	+++	+++	+++	+++	+++
43	VIII. PASSERIFORMES	23. Estrildidae (Waxbills and Allies)	White-rumped Munia (*Lonchura striata*)	R		+	+	+	–	+
44	VIII. PASSERIFORMES	24. Fringillidae (Finches and Allies)	Eurasian Siskin (*Spinus spinus*)	W	#	+	–	–	–	+

(continued)

(continued)

No.	Order	Family	Bird Species	Resident and migrant types	Protection level	Encounter rate(P_i)				
						Spring	Summer	Autumn	Winter	Whole year
45	VIII. PASSERIFORMES	24. Fringillidae (Finches and Allies)	Grey-capped Greenfinch (*Chloris sinica*)	R	#	–	+	–	–	+
46	VIII. PASSERIFORMES	24. Fringillidae (Finches and Allies)	Chinese Grosbeak (*Eophona migratoria*)	W	#	+	+	+	+	+
47	VIII. PASSERIFORMES	25. Emberizidae (Old World Buntings)	Yellow-throated Bunting (*Emberiza elegans*)	W	#	–	–	–	+	+

Note Resident and migrant types[1]: R—Resident; S-Summer Visitor; W-Winter Visitor; P-Passage migrant
Protection level[2]: ◇-Key protected birds in Shanghai;;#-birds with beneficial and has scientific research and economic value
Pi: Pi >10%: "+++"; 10% ≥ Pi > 1%: "++"; Pi ≤ 1%: "+"; non-found: "–"

[1]The classification of resident and migrant bird types refer to:Cai YT, Tang SM, Yuan X, et al. (2011) Shanghai Bird Records and Changes. Journal of Fudan University (Natural Science) 50(3): 334–343

[2]The protection level of birds refer to: Luan XF (2003) Studies on Avian Community of Shanghai and Planning of Conserbation. Ph.D. thesis of East China Normal University, Shanghai